More Cooking Innovations

More Cooking Innovations

Novel Hydrocolloids for Special Dishes

Amos Nussinovitch and Madoka Hirashima

CRC Press
Taylor & Francis Group
Boca Raton London New York

CRC Press is an imprint of the
Taylor & Francis Group, an **informa** business

CRC Press
Taylor & Francis Group
6000 Broken Sound Parkway NW, Suite 300
Boca Raton, FL 33487-2742

Printed on acid-free paper

International Standard Book Number-13: 978-1-138-08409-4 (Hardback)

Visit the Taylor & Francis Website at
http://www.taylorandfrancis.com

and the CRC Press Website at
http://www.crcpress.com

To our teachers and mentors, Prof. Micha Peleg and Prof. Katsuyoshi Nishinari, who served as a lighthouse in a sea of mediocrity and introduced us to the magic and beauty of science.

Contents

PREFACE

Hydrocolloids are among the most commonly used ingredients in the food industry. They function as thickeners, gelling agents, texturizers, stabilizers, and emulsifiers; in addition, they have applications in the areas of edible coatings and flavor release. When manufactured food is reformulated to reduce the fat content, hydrocolloids are mainly used to obtain suitable sensory quality. Hydrocolloids are being increasingly used in the health arena because they provide low-calorie dietary fiber, among many other uses.

Many books describe the different water-soluble polymers (hydrocolloids) and their uses. In 1965, a monograph by M. Glicksman, *Gum Technology in the Food Industry* (Academic Press), presented a technical compilation of information about hydrocolloid technology, pertaining to the food industry. The need for such a book was apparent to most food technologists and scientists, particularly to those engaged in the development of convenience food. This book was followed by three more volumes by Glicksman (1982–1984) entitled *Food Hydrocolloids, volumes I, II, and III* (CRC Press). Volume I was composed of two parts, the first dealing with comparative properties of hydrocolloids and the second with biosynthetic gums. Volume II dealt with natural food exudates and seaweed extracts. Volume III described cellulose gums, plant seed gums, and plant extracts. These books were much more comprehensive than Glicksman's first monograph and were very useful for both food technologists and academics.

In 1982, an excellent book entitled *Handbook of Water-Soluble Gums and Resins* (McGraw Hill Company) was edited by R. L. Davidson.

The book comprised 23 chapters written by advisors and contributors from various universities and industries. It contained information on where water-soluble gums and resins come from, how they are used, how they work, and their individual uses in attaining specific properties and performance. It gave an encyclopedic description of the major commercial varieties of both natural and synthetic gums and resins. Each listing began with a concise overview, followed by full details on the chemistry, properties, handling uses, and other pertinent factors.

In 1997, a monograph by one of us (A. Nussinovitch), entitled *Hydrocolloid Applications, Gum Technology in the Food and Other Industries* (Blackie Academic & Professional), was published. It was comprised of two parts. The first part dealt with brief descriptions of the known hydrocolloids. The second part was devoted to information that was more difficult to locate, namely uses of hydrocolloids in ceramics, cosmetics, and explosives, for glues, for immobilization and encapsulation, in inks and paper, and for the creation of spongy matrices, textiles, and different texturized products. Another monograph by A. Nussinovitch entitled *Water-Soluble Polymer Application in Foods* (Blackwell Science), from 2003, discussed the uses of hydrocolloids in food and biotechnology. It also discussed topics such as hydrocolloid adhesives, hydrocolloid coatings, dry macro- and liquid-core capsules, multilayered products, flavor encapsulation, texturization, cellular solids, and hydrocolloids in the production of special textures. Yet another monograph by A. Nussinovitch from 2010, *Plant Gum Exudates of the World: Sources, Distribution, Properties and Applications* (CRC), provided the most extensive description of plant gum exudates in print. One chapter in that book specifically described the food applications of plant exudates in confectionery, salad dressings and sauces, frozen products, spray-dried products, wine, adhesives, baked products, and beverages, among many other industrial products and animal food.

In 2009, the second edition of *Handbook of Hydrocolloids*, edited by G. O. Phillips and P. A. Williams, was published. This excellent book reviewed over 25 hydrocolloids, covering their structure and properties, processing, functionality, applications, and regulatory status. The book also emphasized protein hydrocolloids and protein–polysaccharide complexes. It expanded the coverage of microbial polysaccharides and discussed the role of hydrocolloids as emulsifiers and dietary fibers.

In 2010, the book *Food Stabilisers, Thickeners and Gelling Agents*, edited by A. Imeson, was published. This practical guide reviewed the incorporation of hydrocolloids in food to give structure, flow, stability, and desirable eating qualities.

These are just a few examples of the wealth of material existing in this field of science. Note that the inclusion of a new book in this short list of published material does not imply that the new work is better than other published books on hydrocolloids.

Although some food recipes can be located in a few of these books, there are no scientific books fully devoted to the fascinating topic of hydrocolloids and their unique applications in the kitchen. A kitchen can be regarded as an experimental laboratory, where food preparation and cookery are done by processes that are well described by the chemical or physical sciences. Finally, it is well established that an understanding of the chemistry and physics of cooking and the involvement of different ingredients (such as hydrocolloids) will lead to improved performance and increased innovation in this realm. Since the use of hydrocolloids is on the rise, a book that covers both past and future uses of hydrocolloids in the kitchen is both timely and of great interest.

In 2014, we authored the book *Cooking Innovations, Using Hydrocolloids for Thickening, Gelling and Emulsification*. That very successful book did not include some important hydrocolloids, among them chitin and chitosan, gum karaya, gum tragacanth, and milk proteins. These are included in the present book. Therefore, this new book completes the work of the first volume, but this is not its only purpose. In addition to the important hydrocolloids that were left out in the first *Cooking Innovations*, we have added chapters about unique hydrocolloids that, in our opinion, will not only be used in future cooking, but will pave the way for new and fascinating recipes and cooking techniques. For example, the recipe for chocolate cookies with arabinoxylan-rich fiber results in a unique and pleasant texture—crunchy and crumbly. Another recipe presents a way to use mesquite meal (with partial inclusion of hydrocolloids), which is utilized in domestic cooking in northwestern Mexico, instead of the pure hydrocolloid (which could be more expensive) to prepare a traditional cornbread. Thus, this book will show the chef or amateur cook that it is possible to look beyond expensive ingredients and perform the job in a less costly way, as some cultures are already doing.

GENERAL APPROACH AND AIMS

Each chapter in this book addresses a particular hydrocolloid, protein hydrocolloid, or protein–polysaccharide complex, in alphabetical order. Each chapter starts with a brief description of the chemical and physical nature of the hydrocolloid, its manufacture, and biological/toxicological properties. It is important to note that this book is not intended as a replacement for the already published books on hydrocolloid properties (some of which are mentioned above); our aim is not to compete with or repeat any of the information found in those books. In the present book, the emphasis is on practical information for professional chefs and amateur cooks alike. Furthermore, such a volume may inspire cooking students, and introduce food technologists to the myriad uses of hydrocolloids—how they are used and for what purposes. Each chapter includes at least one recipe demonstrating the unique properties of the hydrocolloids that can be used in cooking. Several formulations have been chosen for the food technologists, who will be able to manipulate them for large-scale use or use them as a starting point for novel industrial formulations. In summary, the volume is written so that chefs, food engineers, food science students, and other professionals can cull ideas from the recipes and be initiated into the what, where, and why of a particular hydrocolloid's use. Each recipe/formulation is accompanied by color images showing the final food/product. Additional images illustrating some of the more important steps in the cooking/preparation process are also provided with explanations.

The new outline and book will advance both the evolution and revolution in hydrocolloid use in the kitchen today. In 1965, Glicksman's first book about hydrocolloids helped those who were trying to use hydrocolloids in food technology and cooking. Now this book will shed light on some hydrocolloids that were not well-known 50 years ago, but that can produce fantastic recipes and unique results, for example, pullulan's ability to change a food's gloss

CHAPTER 1: HYDROCOLLOIDS—WHERE, WHY, AND WHEN?

Hydrocolloids can be extracted from common or unique natural sources. These include cereal grains, tree wounds, seaweed, animal skin and bones, and fermentation slime, among many others. Aside from those natural sources, synthetic gums are produced by skilled organic chemists. This introductory chapter deals with the terminology and classification of hydrocolloids, their market and economics, gum constituents and their effects on processing, functions of hydrocolloids in food applications, hydrocolloids as viscosity formers and their typical food applications, gelation, gel types and linkages, gel textures, gel-enhancing effects of other gums, the presence of hydrocolloids in emulsions, suspensions and foams, and crystallization control. Other unique food applications of hydrocolloids and regulatory aspects related to their use are also briefly discussed in this chapter.

CHAPTER 2: ARABINOXYLANS

Arabinoxylans are the predominantly non-cellulosic polysaccharides of primary and secondary cell walls, found in cereals and grasses. Aqueous extraction of arabinoxylans from grain and agricultural byproducts provides potentially suitable material for use as gelling agents, cryostabilizers, and as a source for prebiotics. They have important applications in baked goods and dairy products. Therefore, the recipes chosen for this chapter include arabinoxylan soda bread. This recipe is for the traditional, non-leavened (non-yeast) Irish bread that can be eaten as an accompaniment at breakfast. Another recipe is for cake muffins with added arabinoxylan; this recipe can make an American-style muffin in which other ingredients, such as raisins, dried fruit with rum, chocolate and sweet potato, etc., can be added. A recipe for chocolate cookies with arabinoxylan-rich fiber is also presented for their unique and pleasing texture. For variety, it is also possible to use different chocolates in this recipe or change the shape of the biscuits.

CHAPTER 3: BACTERIAL CELLULOSE

Bacterial cellulose is a new functional material with a wide range of applications, which can be attributed to its high purity, crystalline structure, and high water-absorption capacity. Although bacterial cellulose has important applications in a variety of food formulations, its use in the kitchen at this stage is limited. Therefore, we first present a protocol for the commercial preparation of *nata de coco*, and then give recipes for its use as a dessert base. *Nata de coco* is a popular dessert food in the Philippines, which is also exported to Japan. The food is produced by *Gluconacetobacter xylinus*, using coconut milk as the carbon source. The presented recipes based on *nata de coco* include: *nata de coco* and fruit, *nata de coco* in fruit jelly, and milk jelly with *nata de coco*. Another popular bacterial cellulose-containing food product, which can be prepared in the home kitchen, is the Chinese *kombucha* or Manchurian tea. It is obtained by growing a symbiotic culture of the bacteria *Gluconacetobacter* (*Acetobacter*) and yeast in a medium of tea extract and sugar. Another presented recipe is for Russian salad dressing that includes bacterial cellulose. This dressing is good on fresh and boiled vegetables, boiled seafood, or fried fish. Despite its name, this recipe was actually invented in the United States.

CHAPTER 4: CEREAL β-GLUCANS

Cereal β-glucans can be incorporated into a wide range of food formulations, where they impart novel functionalities and improved nutrition. The recipes that we chose for this chapter demonstrate different effects of β-glucans on food properties. In the Indian drink *lassi*, cereal β-glucan is added for its dietary fiber. In the traditional Basque omelet, cereal β-glucan acts as a texture modifier, i.e., it makes the omelet fluffier. In boiled mackerel with vinegar, the cereal β-glucan helps eliminate the fishy smell.

CHAPTER 5: CHITIN AND CHITOSAN

Several grades of chitin and chitosan are commercially available. Chitosan is used as a texturizing agent for perishable food. This chapter includes a commercial preparation recipe for *tempeh*, an Indonesian soy product which contains chitosan. The chapter includes a few examples of products (e.g., croquettes) that use chitin and chitosan as thickeners. In thick fried *tofu*, chitosan is used as a preservative, dietary fiber, and thickening agent. Chitosan added to carrot or apple juice serves as a thickener and a source of fiber.

CHAPTER 6: DEXTRAN

Dextran refers to a class of glucans that are produced extracellularly. The properties of different dextrans depend on their structures. Some dextrans are water-soluble and others are insoluble. In general, dextran has high solubility and low viscosity. Dextrans have the potential for applications in food, as conditioners, stabilizers, bodying agents, and the like. Dextrans are used in confectionery products to enhance moisture retention and viscosity and inhibit sugar crystallization. We present a recipe for dinner rolls, using dextran as a texture modifier; adding dextran results in softer dough with greater volume and a longer shelf life. The very interesting recipe for ice cream cones includes dextran to produce a crisp, frangible, and light texture. These cones can be nested to form a stack of cones without wedging or sticking.

CHAPTER 7: GUM GHATTI

This natural tree gum exudate is a viscosity-former with excellent emulsification properties. In the industry, it is used in beverage emulsification. In the kitchen, it can be used in mayonnaise-type dressings or butter cream. We chose to prepare mayonnaise-type dressing with both gum ghatti and xanthan gum to contribute to its thick consistency. This recipe is especially beneficial for people suffering from egg allergies. A recipe for table syrup with gum ghatti is also presented. This recipe is good for pancakes. Although we prepared it in the laboratory using an experimental apparatus, it can also be made in the kitchen by controlling the temperature by stirring and with the use of everyday cooking utensils.

CHAPTER 8: GUM KARAYA

Gum arabic is by far the most important plant-exudate hydrocolloid. Nevertheless, there are other related gums that have maintained their economic and technological importance over centuries. These include gum tragacanth, gum karaya, gum mesquite, and larchwood arabinogalactan. Gum karaya is the dried exudate of the *Sterculia urens* tree. It is a complex, partially acetylated polysaccharide of high molecular weight. Gum karaya has been used as a stabilizer in ice cream, ice milk, mellorine (a lower-cost imitation ice cream), and related products. Although most homemade sherbets do not contain hydrocolloids, we chose to prepare a fruit sherbet which contained gum karaya and guar gum to prevent crystallization. This formulation was chosen specifically for the benefit of food technologists, who will be able to manipulate it for possible large-scale use.

CHAPTER 9: GUM TRAGACANTH

Gum tragacanth is an exudate. It is employed in numerous low-pH products, such as salad dressings, condiments, and relishes; it also serves as a stabilizer and presents a smooth oral feel by means of its surface-active properties. Gum tragacanth has a wide range of the properties that are required for use in condiments, dressings, and sauces. Gum tragacanth does not degrade at low pH. In addition, it does not require any additional surface-active agent to produce a stable emulsion. Therefore, we present a blue cheese dressing formulation to demonstrate this gum's unique abilities in the kitchen. Another recipe is for a sweet and sour sauce with gum tragacanth.

CHAPTER 10: INULIN

Inulin is the generic name for β-(2,1) fructans. In most cases, inulins are a polydisperse mixture of fructan chains, with a chain-length distribution that depends on the source and time of harvesting. The rheological and sensory characteristics of inulin gels make them an excellent substitute for fat in a wide range of food. One recipe that we chose to prepare was ketchup. This ketchup is easy to make with tomato purée and owes its excellent consistency to the addition of inulin. It is possible to reduce the sugar content in the ketchup a bit more, and adjust the sweet and sour taste by modifying the vinegar and sugar contents. Addition of inulin also eliminates the potentially lumpy texture, typical to reduced-calorie recipes. Another prepared recipe is for low-fat chocolate mousse. The added inulin replaces 50% of the heavy cream used in traditional mousse. Another recipe that includes inulin is *pa-jun* (Korean pancake with green onion). This recipe was chosen due to its versatility with respect to inclusions of vegetables, seafood, and meat. In view of the obesity pandemic, it is estimated that inulin will be included in many future recipes, and its use in the kitchen will rise.

CHAPTER 11: LARCHWOOD ARABINOGALACTAN

Larch arabinogalactan, a plant-derived biopolymer, is an excellent dietary fiber that has been shown to increase the production of short-chain fatty acids, principally butyrate and propionate, and to decrease the generation and absorption of ammonia. Larch arabinogalactan can stimulate natural killer cell cytotoxicity, enhance other functional aspects of the immune system, and inhibit the metastasis of tumor cells to the liver. Its biological source is larch wood. In food, it can be used as an emulsifier, stabilizer, binder, or bonding agent. We chose to present its unique effects on baked goods (sugar snap cookies); it improves the texture and reduces the stickiness of the dough.

CHAPTER 12: LEVAN

Levan is an extracellular polysaccharide. It can be produced by a variety of microorganisms and plants. Levan does not swell in water and has very low intrinsic viscosity. It can be used as an emulsifying, stabilizing, or sweetening agent in food. It is utilized for the production of edible food coatings. Levan can be effectively thermally processed and we demonstrate this property in a recipe where levan is added to walnut meringues to strengthen the foam's stability and improve its texture.

CHAPTER 13: MESQUITE GUM

Mesquite gum is an exudate produced from the bark of *Prosopis* spp. Mesquite gum has been used extensively in a variety of food applications. The gum is not yet produced on a large scale, and due to the presence of tannins, it is not permitted as a food additive in the United States. However, mesquite meal is used in domestic cooking in the Sonora region of northwestern Mexico, and the use of mesquite meal in baked goods is demonstrated.

CHAPTER 14: MILK PROTEINS

Milk proteins can be used in baked goods, dairy products, beverages, dessert-type food, pasta products, confectionery, meat, and textured products. In this chapter, we present a few common recipes, for cottage cheese and whey, and vanilla ice cream. More interesting, however, is the use of casein in baked products (bread). We chose to present a recipe for egg pasta in which whey protein concentrate powder was included to prepare a protein-rich product, and a recipe for cocoa cookies that includes milk protein concentrate to produce both unique hard-type cookies and a protein-rich food.

CHAPTER 15: OTHER MICROBIAL POLYSACCHARIDES: ALTERNAN, ELSINAN, AND SCLEROGLUCAN

The major microbial polysaccharides are xanthan gum, curdlan, gellan gum, and bacterial cellulose. However, there are several other microbial polysaccharides (i.e., alternan, elsinan, and scleroglucan) that possess important or potential applications in the food industry. Although today they are rarely found in the kitchen, different recipes are presented to demonstrate the uniqueness of these polysaccharides and their parts. We chose to prepare gluten-free bread due to its importance for those suffering from gluten-related disorders.

CHAPTER 16: PULLULAN

Pullulan is an extracellular polysaccharide. It can be produced via fermentation by *Aureobasidium pullulans*. Pullulan has special uses in food. It is regarded as a dietary fiber and it promotes the growth of beneficial bifidobacteria. Pullulan can be used as a substitute for starch in pastas or baked goods. Other possible uses are as fillers in beverages and sauces. We chose to use pullulan as a texturizer of almond cookies; the pullulan in this recipe was used as a binding and water-retention agent. In a recipe for *teriyaki* sauce, pullulan was used as an adhesive and gloss agent.

CHAPTER 17: SOLUBLE SOYBEAN POLYSACCHARIDE

Soluble soybean polysaccharide (SSPS) is extracted and refined from soybean. Its functional properties and applications are numerous.

In Japan, SSPS is classified as both a food ingredient and a food additive, with no limitations on its application. A recipe for a yogurt drink with the addition of SSPS is presented to emphasize its stabilization ability in low pH environments, as well as its foam-stabilizing ability. This ability is beneficial in other products as well. Boiled pasta with SSPS is included as an example of food that will maintain its textural properties when retort-packed or frozen. We also include a recipe for steamed rice in which SSPS is used to make it non-sticky, as well as to enhance the water-holding capacity of the product.

CHAPTER 18: VEGETABLE PROTEIN ISOLATES

The main vegetable protein sources are legumes, cereals, oilseeds, roots, and green leaves. Food applications for seeds include baked goods, dessert sauces and dairy substitutes, infant formulas, fortified food, meat and sausage substitutes, beverages, soups and gravies, and cheese-like products. A variety of different recipes based on various raw materials are included in this chapter. These include *momen tofu* (soybean curd), soybean milk and black sesame pudding, *unohana* (seasoned *okara, or soybean fiber*), low-calorie *okara* pound cake, basil and sunflower seed paste, lotus root balls, alfalfa stew, and fruit juice with alfalfa.

CHAPTER 19: XYLOGLUCAN

Xyloglucan is a major structural polysaccharide found in the primary cell wall of higher plants.

Tamarind seed xyloglucan (TSX) is widely used as a food additive in Japan and Southeast Asia (i.e., Taiwan, South Korea, and China). TSX is used mainly in sauces, dressings and mayonnaise, ice cream and some flour products. A recipe for sponge cake demonstrates the use of TSX as a texturizer and an inhibitor of starch retrogradation; in *tsukudani* (laver preserves), TSX enhances water-holding capacity and serves as a thickener. The gelling ability of TSX is demonstrated in the recipe for *kudzu-mochi* (Japanese arrowroot jelly). Here, TSX is also used as an inhibitory agent of starch retrogradation. Another interesting recipe is for sweet red bean soup (*shiruko*) where TSX is added to stabilize the suspension.

CHAPTER 20: FUTURE IDEAS FOR HYDROCOLLOID PROCESSING AND COOKING

There are many advantages of hydrocolloids that will most certainly contribute to their increased use in food and cooking in the future. In this chapter, we discuss the future trends of several gelling agents and viscosity-formers, as well as the future trends in protein hydrocolloids. The unique nutritional and future health claims of hydrocolloids, as dietary fibers, are discussed. Some related issues as well as the topic of hydrocolloids alternatives are also discussed.

This book was written by two authors who are entranced by the field of hydrocolloids, their potential applications, and cooking. This book was written with the intention of creating something much more than just a traditional cookbook. The information provided here on the various hydrocolloids is useful and extensive, but it is not strictly theoretical. We have made an attempt to include recipes that everyone can follow within the confines of their own kitchen—be they professional chefs or amateur cooks. This book also contains discussions on products and scientific experiments that can be conducted and studied in university laboratories, like the ones that are currently being performed by one of the authors (M. Hirashima) in Japan. As such, the book can be used as a textbook for cooking science and food-processing classes. This book provides recipes that can be scaled up for industrial use, thus making it ideal for food technologists and engineers. We believe that such a book can bridge the gap between the scientist and the chef or, in fact, anyone who is interested in novel applications and textures in the kitchen. In addition, it is designed to serve as a guide for all those who want to introduce the fascinating world of hydrocolloids to the public.

We believe that this is going to be an extremely useful book. It can be used by personnel who are involved in food formulation, food science, and food technology. In particular, it will be helpful for: food scientists (chemists/microbiologists/technologists) working in product development; food engineers whose job typically involves figuring out how to make the products that the developers cook up; and research chefs, e.g., members of the Research Chefs Association, professional restaurant chefs who like to experiment with new creations, members of the American Culinary Federation and culinary education programs, and members of equivalent organizations in foreign countries. This book will be a useful addition to the traditional libraries of universities and research institutes where food science, chemistry, life sciences, and other practical industrial issues are taught and studied. In this sense, the book is quite unique, and we are confident that it will be a great success.

Amos Nussinovitch and Madoka Hirashima

ACKNOWLEDGMENT

This book was written over the course of 2 years, following the success of our previous book: *Cooking Innovations, Using Hydrocolloids for Thickening, Gelling, and Emulsification*. It describes more cooking innovations, in particular, the use of novel hydrocolloids in special dishes. The book is fully devoted to the fascinating topic of hydrocolloids and their unique applications in the kitchen. This book was written with the belief that an understanding of the chemistry and physics of cooking and the involvement of hydrocolloids will lead to improved performance and increased innovation in this realm. The use of hydrocolloids is on the rise in numerous fields. The time is therefore ripe for this book, which covers both past and future uses of hydrocolloids in the kitchen. We have tried to include many traditional and non-traditional, local and international uses of hydrocolloids in cooking. We hope this book will assist readers in gaining comprehensive knowledge about hydrocolloids, and help them stay updated on past and current uses and applications of hydrocolloids in cooking. The volume is written so that chefs, amateur cooks, food engineers, food science students, and other professionals can cull ideas from the recipes and be initiated into the what, where, and why of a particular hydrocolloid's use.

We wish to thank the publishers for giving us the opportunity to write this book. Special thanks to Stephen Zollo, the senior editor of Food Science & Technology at CRC Press, Taylor and Francis Group, for his efficient handling of the project, from its conception to the moment we received the green light to start cooking, photographing, and writing. Stephen's genuine interest, enthusiasm, and encouragement during the process were phenomenal and are deeply appreciated. We wish to thank our editor, Camille Vainstein, for working shoulder-to-shoulder with us when time was getting short. We are grateful to Gal Sason and to the illustrator, Lotem Sason, for drawing the wonderful cover art for this book.

We wish to thank Daiwa Pharmaceutical Company for providing us with arabinoxylan samples, Gan Shmuel Group for providing us with a levan sample, Hayashibara Company for pullulan samples, and Fuji Oil Company for soluble soybean polysaccharides. We are also thankful to Ms. Ayumi Shibamura, from DSP Gokyo Food & Chemical Company, for supplying us with xyloglucan samples, Dr. Makoto Nakauma, from San-Ei-Gen F. F. I., for bacterial cellulose samples, and Dr. Kobi Meiri, from the Tnuva Company, Israel, for contributing the different milk proteins.

Last, but not least, we wish to thank the Hebrew University of Jerusalem and Mie University for being our home and refuge for many years of extensive research and teaching.

Amos Nussinovitch and Madoka Hirashima
Israel and Japan

AUTHORS

Professor Amos Nussinovitch was born in Kibbutz Megiddo, Israel. He studied Chemistry at the University of Tel Aviv, and Food Engineering and Biotechnology at the Technion – Israel Institute of Technology. He has worked as an engineer at several companies and has been involved in a number of R&D projects, in both the United States and Israel, focusing on the mechanical properties of liquids, semi-solids, solids, and powders. He is currently at the Biochemistry and Food Science Department of the Robert H. Smith Faculty of Agriculture, Food and Environment of the Hebrew University of Jerusalem, where he leads a large group of researchers working on theoretical and practical aspects of hydrocolloids. Prof. Nussinovitch is the sole author of the books: *Hydrocolloid Applications*, *Gum Technology in the Food and Other Industries*; *Water-Soluble Polymer Applications in Foods*; *Plant Gum Exudates of the World: Sources, Distribution, Properties, and Applications*; *Polymer Macro- and Micro-Gel Beads–Fundamentals and Applications; Adhesion in Foods, Fundamental Principles and Applications.* He and his present co-author Dr. Madoka Hirashima (see below) recently co-authored the book: *Cooking Innovations, Using Hydrocolloids for Thickening, Gelling, and Emulsification.* Prof. Nussinovitch is the author or co-author of numerous papers on hydrocolloids and on the physical properties of food, and he has many patents. This book is devoted specifically to more cooking innovations, specifically, using traditional and numerous novel hydrocolloids for special dishes. The author has been working in this area for many years and has studied gel textures and structures, textures of hydrocolloid beads and texturized fruit, liquid-core hydrocolloid capsules, different hydrocolloid carriers for encapsulation, novel hydrocolloid cellular solids and edible hydrocolloid coatings of food, among many other applications. Several years ago, Prof. Nussinovitch received a lifetime award from the Manufacturers Association of Israel for his unique and considerable contributions to both academia and the food industry in Israel.

Madoka Hirashima, Ph.D., was born in Kyoto, Japan. She studied the rheological properties of curdlan and cornstarch at the Graduate School of Human Life Science, Osaka City University. She worked at a food company as a novel food developer, and then as a lecturer at several colleges. She is currently in Home Economics Education at the Faculty of Education, Mie University, where she teaches cooking as well as cooking science. She continues to study the rheological properties of polysaccharides, with a focus on the textures of starch and *konjac* products.

CHAPTER 1

Novel Hydrocolloids—Where, Why, and When?

1.1 INTRODUCTION

Hydrocolloids can be extracted from common or from exceptional natural sources, such as animal bones and skin, cereal grains, fermentation slime, seaweed, and wounded trees, among others. In addition to these natural sources, synthetic gums can be manufactured by trained organic chemists. Emerging microbial polysaccharides, e.g., pullulan (Chapter 16), scleroglucan, elsinan, alternan (Chapter 15), and levan (Chapter 12), have recently begun moving out of the novel scientific territory and are being put into applied use. Further progress includes the introduction of hydrocolloid hybrids which have the potential to deliver innovative functionalities. Another example is combinations of hydrocolloids that generally impart innovative and enhanced rheological characteristics to food products and reduce costs. The available hydrocolloids are either used as food or are used in food products; besides that, gums are also employed for non-food purposes. The importance of hydrophilic colloids lies in their hydrophilic nature, which makes gums the essential textural constituents of most foods.

1.2 TERMINOLOGY

The classification of water-soluble gums is inconsistent, possibly owing to its casual development over numerous centuries. Indeed, this inconsistency reveals an accumulation of terminologies from various geographical sources applied to a diversity of impure natural substances. Initially, the word "gum" was most probably used to designate the natural gum exudates oozing from trees. This term must have referred to every type of natural exudate, encompassing various water-insoluble materials, for example, resins, latex, and chicle. This practice would explain the incorrect usage of the word "gum" for many water-insoluble resins as well, which are used in the paint and chemical industries today. The development of trade led to an abundance of vague common names and trade names.

In 1947, Mantel specified the limits of the expression "water-soluble gums". These materials do not dissolve in water in a precise scientific sense, but form colloidal dispersions. In 1953, Whistler and Smart proposed the generic phrase "glycan" to designate a "polysaccharide". The suffix "-ose", which refers to the basic sugar in a polymer (glucose), was substituted by the suffix "-an" to develop this new phrase, which was then repeatedly found in names linked to sugar polymers. Publications from 1959 onwards made an attempt to bring further uniformity in this field. For example, rather than using the term "Danish agar" for gum extracted from *Furcellaria fastigiata*, they chose to use the term "furcellaran". Larchwood gum (Chapter 11) was frequently termed as "arabinogalactan" and was sold under the trade name "Stractan"—both names are in compliance with the "-an" suffix. Glicksman proposed that a gum be defined as any polymeric material that can be dissolved or dispersed in water to obtain a viscous solution or dispersion. This classification was based on distinguishing the well-designed properties of gums as used in the trade. Furthermore, the classification included synthetic polymers and proteins, such as gelatin and casein (Chapter 14), which display viscosity- and/or gel-forming properties in water. This characterization differentiated the water-soluble polymers from oil-soluble resins (which are also called gums), such as gum kauri and gum copal. It also distinguished them from other polymers, for instance, rubber and gum chicle, that are incorrectly designated as gums.

1.3 CLASSIFICATION

By tradition, most of the gums were regarded as polysaccharides, and they were clustered into the same group, in keeping with their plant source. Therefore, agar–agar, algin, and carrageenan were brought together in the seaweed group. Gum arabic, gum karaya (Chapter 8), and gum tragacanth (Chapter 9), among others, were categorized in the tree exudate group. Other gum-like substances, for example, starch and pectin, were treated as distinct groups, whereas proteins, such as gelatin, were not included in any group. Moreover, there was no room for synthetic gums, such as cellulose derivatives and vinyl polymers, which required a different category. One possible approach to classification could be the use of botanical origin to categorize central plant gums. For instance, locust bean gum (LBG) and guar gum are derived from similar plant seed sources and have similarities in their chemical structures; in addition, they can sometimes be employed for analogous purposes. To allow wide-range sorting, all varieties of gums that are used in the food industry would be contained within that category, and room would be left for new gums that might be established in the future. As a consequence, Glicksman suggested an all-inclusive grouping composed of three categories: gums found in nature, modified gums (semisynthetic) that are based on chemical modifications of natural gums, and synthetic gums that are manufactured by chemical synthesis.

1.4 ECONOMICS

The global hydrocolloid market reached a value of $6.6 billion in 2015. Growing at an estimated 5-year (2015–2020) compound annual growth rate (CAGR) of 4.4%, the market should reach 2.1 million metric tons in 2020, with a value of almost $8.2 billion. Europe is expected to continue to be the largest market, with a market share of 33.5% in terms of value. Nevertheless, by 2020, the Asia-Pacific region is predicted to become the second-largest market, with a market share of 27%, followed by North America at 23.9%. The increasing demand for hydrocolloids in, for example, the food and beverage industries, oil drilling, pharmaceuticals, nutraceuticals, cosmetics, textile printing, and paper treatment, combined with the growth of these industries, should provide for a modest market growth, particularly in the Asia-Pacific, European, and North American regions. Additional key market drivers are factors like rising population and growing concerns for food safety and quality among consumers.

1.5 GUM CONSTITUENTS AND THEIR EFFECTS ON PROCESSING

Gum constituents are found in just about every natural food, frequently accounting for the structural and textural properties of the source plant. In manufactured foods, hydrocolloids impart textural and functional properties in the form of food additives. Furthermore, convenience foods almost always include hydrocolloids in their list of ingredients. The use of gums to produce higher quality food can be exemplified by ice cream; homemade ice cream commonly has poor textural qualities, like the presence of ice crystals, a sandy texture, and an absence of smooth meltdown. Today's technologically manufactured ice creams contain multiple hydrocolloids as stabilizers and emulsifiers to eliminate these quality defects. This practice is prevalent in all food processing systems that involve changes in moisture content or the water's physical shape. Together, the residual gum constituents and the added hydrocolloids govern the type of physical transformation and the rate of migration of the water component. As a whole, the added gums might help in handling processing settings via alterations in water retention, decreasing evaporation rates, modifications in freezing rates, modifications in ice-crystal formation, and involvement in chemical reactions. These applied effects are established in the textural qualities of the items for consumption. The useful effects are taken into account along with aspects such as price, convenience, simplicity of usage, and legal restrictions related to the hydrocolloid's use.

1.6 FUNCTIONS OF HYDROCOLLOIDS IN FOOD APPLICATIONS

1.6.1 Functions in Food Products

Hydrocolloids have a wide array of applications. A few examples are as adhesives in bakery glazes; binding agents in sausages; bulking agents in dietetic foods; crystallization inhibitors in ice cream and sugar syrups; clarifying agents in beer and wine; clouding agents in fruit juices; coating agents in confectionery; emulsifiers in salad dressings; encapsulating agents for powdered fixed flavors; film formers for sausage casings and additional protective coatings; flocculating agents in wine; foam stabilizers in whipped toppings and beer; gelling agents in puddings, desserts, aspics, and mousses; mold-release agents for gum drops and jelly candies; protective colloids in flavor emulsions; stabilizers in beer and mayonnaise; suspending agents in chocolate milk; swelling agents in processed meats; syneresis inhibitors in cheese and frozen foods; thickening agents in jams, pie fillings, sauces, and gravies; and whipping agents in toppings and icings. It is important to note that the comprehensive applications of gums are confined to their two central properties—to serve as thickening and gelling agents. The ability to increase viscosity, i.e., viscosity production, is a crucial feature in the use of hydrocolloids as bodying, stabilizing, and emulsifying agents in foods. A few gums that have gelling abilities are used as a support in foods where shape retention is desirable to counter the application of pressure. The most recognizable gelled food item is the gelatin dessert gel; other recognized food gels are starch-based milk puddings, gelatin aspics, and pectin-gelled cranberry sauce.

1.6.2 Viscosity Formation and Its Typical Food Applications

Despite the fact that hydrocolloids are exploited to impart viscosity to food items, the foremost concern is product stability. This can be controlled by a suitable hydrocolloid selection. It is vital to avoid degradation of the polymer, which would result in decreased viscosity and consequent product deterioration. Almost all gums are long-chain polymers; as a consequence of their molecular breakdown, their viscosities have a tendency to decrease in solution. Such degradation can be the outcome of high shearing and/or processing temperatures. Natural, low-viscosity gums are more stable than the high-viscosity variants. Thus, stability comparisons should be made based on equal viscosities rather than equal concentrations. Viscosities of hydrocolloid systems depend on concentration, temperature, degree of dispersion, solvation, electrical charge, previous thermal and mechanical treatments, and the presence or lack of electrolytes and non-electrolytes. Gums are often used exclusively for thickening in food items, for instance, in dry beverage mixes and pie fillings. In the latter, if the filling is fruit-based, then gums are used to increase the viscosity of the fruit juice, so that the filling flows slowly once it is in the pie shell. In the U.S. Europe, and Latin America, agar is used in pie fillings. Adding xanthan gum to either cold- or hot-processed baked goods and fruit pie fillings improves their texture and flavor release. Carob bean gum is used in pie filling as thickening agent. In addition to their functions as emulsifiers and nutritional enhancers, milk proteins (Chapter 14) can also improve the texture of pie fillings. In beverages, gums are used to generate body, predominantly in sugar-free dietetic drinks. Dry mixes of xanthan and maltodextrin are combined for direct thickening of beverages, such as water, fruit juice, green tea, and milk, that are sold in Japan to elderly people who have difficulty swallowing. Milk protein products are used as stabilizers with plentiful applications in beverages. Vegetable protein isolates (Chapter 18) are used to affect solubility, viscosity, and acid stability in a variety of beverages. Soybean applications include dietetic and health foods, beverages, and dairy analogues. Soluble soybean polysaccharides (Chapter 17) applied to beverages provide excellent stability and a refreshing taste under acidic conditions. Larchwood Arabinogalactan (Chapter 11) is highly water-soluble and, thus, freely disperses in hot beverages within 30 seconds and does not cause turbidity or changes in viscosity. Mesquite gum (Chapter 13) is used in foods and beverages in Mexico. At a very low concentration (0.05%), xyloglucan (Chapter 19) can enhance the body or improve the texture of low-fat milk and fruit beverages. Pullulan (Chapter 16) produces fairly low-viscosity solutions and is, thus, used as a low-viscosity filler in beverages and sauces. Cereal β-glucans (Chapter 4) are employed in cholesterol-lowering beverages. Standardization of stabilized beverage systems that contain thickeners (xanthan, guar gum, and pectins) is achieved by adding inulin (Chapter 10).

In soup mixes or prepared soups, body and consistency are obtained with starches and supplementary hydrocolloids. Similarly, sauces and sauce blends contain gums to obtain the desired consistency. Whey solids (Chapter 14) are added to dehydrated soup mixes and sauces to impart a dairy flavor, to enhance additional flavors, and to provide emulsifying and stabilizing effects. Caseinates (Chapter 14) are used as emulsifying agents and to control viscosity in canned cream soups and sauces, and in the manufacture of dry emulsions for use in dehydrated cream soups and sauces. If low-pH sauces are manufactured, gum tragacanth (Chapter 9) and propylene glycol alginate can offer resistance to acid degradation. In South Africa, gum ghatti (Chapter 7) is acceptable as an emulsifier in condiments and sauces. Gum karaya (Chapter 8) is also used in sauces and dressings. Tamarind seed xyloglucan (TSX) (Chapter 19) is used as a common food additive in Japan. TSX is chiefly used in sauces, dressings and mayonnaise, ice cream, and some flour products. Bacterial cellulose (Chapter 3) can be possibly used in pourable and spoonable dressings, sauces, and gravies. Today, gums are finding novel uses in dried pet foods. In dry dog food, addition of water can hydrate and thicken the meat-like pieces, giving a thick gravy-like sauce surrounding meat-like chunks.

1.6.3 Gelation, Gel Types, and Linkages

Gels obtained from gums can keep their firm structural form under pressure. Gels have a continuous network of solid material encompassing an uninterrupted or finely separated liquid phase. The solid material frequently consists of long-chain molecules in a structure of fibrils interlinked by primary or secondary bonds at various locations along the molecule. Gels display properties of both solids and liquids. The similarity to solids is reflected in their structural stiffness and characteristic elastic response, and to liquids in their compressibility, electrical conductivity, and vapor pressure. Gelation starts with a continuous decrease in the Brownian movement of colloidal particles occluded within the gel. This decrease is initiated by the application of long-range forces on the molecules, which results in hydration and coherence of the particles. An increase in viscosity parallels the proceeding gelation; the fluid is gradually absorbed by the swelling solid, and thus, it is slowly immobilized. As gelation progresses, a 3D network that contains the liquid is gradually developed. Further progress creates a large continuous structure with apparent rigidity. Segments of the large molecular chains in the network can also react with other parts by cross-linking to further increase the rigidity of the entire structure. Gelation can be induced from either the sol or solid-state condition. In the sol state, the gel can be structured by generating forces between the solute molecules; this can be achieved by adding a non-solvent, evaporating the system's concurrent solvent, adding a cross-linking agent, reducing solute solubility by chemical reaction, changing the temperature, or adjusting the pH. In the solid state, gels are formed when adequate liquid is imbibed by the solid phase. In this case, the gel is considered to be in a transitional state of hydration, i.e., between sol and solid. The gel is composed of a continuous network that entraps the liquid phase containing solvent and solutes, some of them comprising non-cross-linked polymeric materials. Cross-linking mechanisms include hydrogen bonding, electrovalent linkage, and direct covalent linkage.

Innovative manufacturing techniques include the construction of sheared gels that display pioneering rheological characteristics. This involves applying shear as the hydrocolloid is undergoing gelation, which characteristically results in the formation of micron-size hydrocolloid gel particles. At an effectively high concentration, the formed structures exhibit strong shear-thinning features.

1.6.4 Gel Textures

Gel textures differ extensively, from smooth, elastic gelatin–water gels to brittle carrageenan–water gels. In gelatin gels, elasticity is attained when the long-chain molecules have easily rotating links; in this case, there are weak secondary forces among the molecules that are interlocked in a few places along their length to form a 3D network. Elastic gel properties prevail in a polymeric system with a low degree of cross-linking; a more rigid gel structure is achieved when a high degree of closely spaced cross-linking exists. Brittle gels can also be achieved by some form of precipitation rather than true gel formation. Such gels might be formed from poorly formulated pectin and alginate gels cross-linked with calcium ions. In some gels, when constituent particles clot after coming into close contact and then shrink, fluid is exuded from the gel; this phenomenon is termed syneresis.

The use of proteins in the manufacture of simulated meat products is an important food application. A chewy gel is required for such products. Chewy gels can be obtained under specific conditions of protein concentration, pH, and temperature. The foremost concern is to produce gels that have sufficient water to be pleasingly moist and, yet, have satisfactory firmness to provide just the "right" resistance to bite. When soy proteins were used to form simulated meat fibers, the desired texture was achieved by changing the pH. More often than not, food gels contain substantial quantities of sugar, which contribute to the flavor and also have a vital functional effect on the gel. Sugar acts as a plasticizer to allow greater separation of the polymer chains, and it also competes with the polymer for water; in other words, sugar decreases the solubility of the polymer. These effects can be complementary when they intensify the elastic properties of the gel or increase the concentration of the cross-linkages, resulting in a broken gel structure.

1.6.5 Gel-Enhancing Effects of Other Gums

Addition of non-gelling hydrocolloids to systems that contain gelling agents can produce unique textures. Synergy occurs when carrageenan and LBG are combined. Carrageenan gels have a brittle, crumbly texture. With the addition of LBG, they become elastic and tender, and are sometimes stronger than gels containing only carrageenan. A similar effect is observed when LBG is added to pectin or agar gels. Perhaps the best-known synergy is exhibited by carrageenan and milk proteins (Chapter 14). Some of the first uses of carrageenan were in milk gels and flans, and in the stabilization of evaporated milk and ice cream mixes; the synergy was possible even with levels as low as 0.03%. The texture of products such as meringues, pie fillings, and marshmallows can be reformed by the addition of non-gelling hydrocolloids. An example is the addition of gum arabic to agar gels to soften their texture. Use of whey proteins (Chapter 14) as a replacement for egg white in the production of meringues formed an adequate product only when defatted ingredients were used. In the production of agar-based marshmallows, addition of gum arabic to the composition produced a tender consistency. With pectin gels, addition of carboxymethylcellulose (CMC) or sodium alginate imparts a smooth texture to the product. Addition of CMC to characteristic alginate dessert gels enhanced their resistance to textural deterioration as a consequence of freezing and thawing cycles. Inulin (Chapter 10) and oligofructose can restore the anticipated sensory attributes to low-fat and low-sugar ice cream products; in this case, inulin contributes to a creamy mouthfeel, consistent melting, and enhanced heat-shock stability, while oligofructose provides a sweet taste, scoopability, and synergy with high-intensity sweeteners, thereby masking any aftertaste.

1.6.6 Hydrocolloids in Emulsions, Suspensions, and Foams and in Crystallization Control

In most situations, hydrocolloids act as emulsion stabilizers by increasing the viscosity of the aqueous phase so that it approaches, and sometimes surpasses, the viscosity of the oil phase. In this process, the tendency of the dispersed phase to coalesce is reduced, and thus, the emulsion is stabilized. For instance, in the case of French salad dressing, i.e., an oil-in-water emulsion, gum tragacanth (Chapter 9) and propylene glycol alginate can be used to stabilize the emulsion. Other salad dressings can be stabilized with starches and whey protein (Chapter 14) products, which can replace the egg yolk in salad dressings. Gum karaya (Chapter 8) is a good emulsion stabilizer for French-style salad dressings since it increases the viscosity of the aqueous phase of the oil-in-water emulsion. Because TSX (Chapter 19) supports emulsion stability, it has been used in salad dressings and mayonnaise. The addition of minute amounts of supplementary gums to starch will also improve the body and shelf life of the product. Hydrocolloid stabilizers can be utilized to stabilize additional emulsions. In these cases, they serve to simulate smoothness, enhance body and meltdown, and improve resistance to heat shock. In numerous applications, the use of a mixture of a few gums is favored owing to their synergistic effects. In its powdery form, gum tragacanth (Chapter 9) can also be used as a carrier for medicines. For instance, gum tragacanth was engaged as a vehicle for an advanced non-nucleoside reverse transcriptase inhibitor of human immunodeficiency virus type 1 (HIV-1). The inhibitor was suspended in 0.5% gum tragacanth, and this suspension was administered orally to dogs, male rats, and monkeys at various concentrations of the enclosed compound; its absorption pattern was studied to understand its distribution to the brain, effects on hepatic drug-metabolizing enzymes, and biliary excretion. Gum tragacanth is also used for the enhancement of stable emulsions containing 50% insect repellant. These were effective as pure repellant compounds against ants, chiggers, mites, mosquitoes, and some fleas.

A stable suspension of solids in a liquid phase can be achieved in chocolate-flavored products. For example, in chocolate milk or syrups, the cocoa solids are suspended by increasing the viscosity of the liquid phase. Carrageenans are suitable for this purpose, because they can concomitantly raise viscosity and react directly with small amounts of protein to form stable colloidal suspensions. Arabinogalactan (Chapter 11) has been established as an innovative protective agent that maintains precious metal nanoparticles in colloidal suspension. Other products containing milk protein solids (Chapter 14), such as buttermilk and yogurt, can profit from effective suspension ability. Hydrocolloids are used to coat particles and change their surface properties. The coating has an affinity for the continuous phase and thus attaches the phases and stabilizes the suspension. Occasionally, surfactants are also involved to better displace the air on the particles with the protective gum. Furthermore, plasticizers, such as glycerol and propylene glycol, are added to obtain a smooth spread.

The construction of stable foams can be improved by adding different proteins. Egg albumin, gelatin, milk (Chapter 14), and soy proteins can be used in confectionery foams. The proteins are dispersed in a sugar solution, and the suspension is whipped. The foam is stabilized by heat-denaturing the egg white protein. If gelatin is used, a stable foam is formed upon cooling. Chemical means can also be used for such purposes. Hydrocolloids, such as sodium alginate, carrageenan, and

LBG, can be added to react with the protein and produce stable foams. In the case of nougat confections, sugar crystals form a grainy structure that holds the whipped egg or vegetable protein foam in its delicate state, thus preventing its breakdown. Icings (i.e., aerated protein systems) on baked goods can be stabilized by thermostable gel stabilizers, such as agar, carrageenan, or gelatin. If the method does not consist of a heating stage, then CMC and cold-water-soluble carrageenans are used. Addition of LBG, carrageenan, or gum karaya (Chapter 8) is useful for whipped toppings that include 20% butterfat. If the butterfat is replaced with vegetable oil and non-milk proteins, then an emulsifier is required to stabilize the dispersed fat and a stabilizer, for example, carrageenan, is added to stabilize the foam. When stable citrus juice foams are required, modified soy albumin can be used as a whipping agent in addition to methylcellulose, which is used as a stabilizer.

Hydrocolloids can affect crystallization by three mechanisms: (i) the hydrocolloid can attach itself to a growing crystal surface via hydrogen bonds or negative- and positive-charge bonding; (ii) the hydrocolloid can compete for the crystal's building blocks. For instance, hydrocolloids will compete with ice crystals for water molecules; (iii) the hydrocolloid can combine with impurities that affect crystal growth. In foods, the main crystalline materials are ice (water) and sugar. Good-quality ice cream, for example, has ice crystals of minimal size.

1.6.7 Other Unique Food Applications

A competent flavor-fixation process involves molecular absorption, where flavor molecules are attracted to the hydrocolloid molecules (gum arabic). The porous gum particles take in and hold the volatile liquid flavors and protect them. The manufacturing process includes spray-drying. Additional gums may also be well suited to such an operation. Powdered flavors are widely used in the food industry. Aside from the gum, the encapsulating process must also take into account numerous other factors, like the percentage of solids in the initial formulation, the ratio of gum-to-oil, the choice of an appropriate emulsifier, the viscosity of the preparation, and drying temperatures. Today, a large number of milk protein products (Chapter 14) are recovered from milk, and it is quite likely that this range will be further expanded in the future. In the continuing search for new milk-derived peptides, the focus might shift to the impact of interactions with other food constituents and technological processing procedures on the biological activities of hydrocolloids. Milk proteins are also used as emulsifying and fat-encapsulating agents in the production of high-fat powders that are used as shortening agents in baking or cooking. Spray-dried gum-based products suffer from several drawbacks: the exposed surface of the particles is large; in gelatin media, the addition of a non-suitable gum at some concentrations can cause unattractive cloudiness. Moreover, in cold-water beverages, the gum is more gradually dissolved, and in some cases, the smallest particles of gum-flavor powder tend to aggregate and settle in the package. Aside from spray-drying, flavors are also fixed in gelatin and the formed slabs of gelled material are broken into fine particles. Such slab-fixation is used in chewing gums and desserts. Supplementary techniques for fixing flavors are freeze-drying and drum-drying. Microencapsulation of flavor oils has also been performed with gum arabic and gelatin using the basic technique of co-acervation. An alternate technique is the encapsulation of flavor oils within insoluble calcium-alginate films.

The genetic modification of cereal β-glucan (Chapter 4) structure may increase its potential human health benefits. In addition, with an uncertain future for consumer acceptability of genetically modified food products, plant breeding can also play an important role in manufacturing cereal glucans with higher yields, novel structures and broader functionality.

The use of edible protective films prepared from various hydrocolloids is gaining in popularity. Alginate films have been used to coat meat and fish. Other edible films have been manufactured from gelatin, pectin, and CMC, among many others. These films are thin and inexpensive, and can be used not only to coat fresh produce for shelf-life extension but also to produce pouches for soluble coffee, tea, soup, and sugar. Such coatings can also be used as wrapping for sticky candies and confections. Another function of gums in food processing is the prevention or minimization of syneresis. It is conceivable that gum addition will be advantageous not only in preventing syneresis but also in refining the texture of the treated products. Such an effect has been achieved in processed cheeses and cheese spreads.

1.7 REGULATORY ASPECTS

Hydrocolloids are regulated as either food additives or food ingredients. In every chapter of this book, the relevant hydrocolloid's regulatory status is noted. Food additives are certified only when their addition can be justified by the user, and not the supplier or manufacturer, in terms of rational technological needs. Furthermore, additives can be used only if they do not present health hazards at any level, and the consumer is not misled in the process.

REFERENCES AND FURTHER READING

Davidson, R. L. 1980. *Handbook of water-soluble gums and resins*. New York: McGraw-Hill.

Dickinson, E., and P. Walstra. 1993. *Food colloids and polymers, stability and mechanical properties*. Cambridge: Royal Society of Chemistry.

Doxastakis, G., and V. Kiosseoglou. 2000. *Novel macromolecules in food systems*. Amsterdam: Elsevier Science.

Glicksman, M. 1969. *Gum technology in the food industry*. New York and London: Academic Press.

Glicksman, M. 1983. *Food hydrocolloids*, vols. 1, 2 & 3. Boca Raton: CRC Press Inc.

Harris, P. 1990. *Food gels*. New York: Elsevier Science Publishing Co.

Hoefler, A. C. 2004. *Hydrocolloids: Practical guides for the food industry*. St. Paul, MN: American Association of Cereal Chemists.

Hollingworth, C. S. 2010. *Food hydrocolloids: Characteristics, properties and structures*. New York: Nova Science Publishers, Inc.

Howes, F. N. 1949. *Vegetable gums and resins*. Waltham, MA: The Chronica Botanica Co.

Imeson, A. 1992. *Thickeners and gelling agents for food*. London: Blackie Academic and Professional.

Laaman, T. R. 2010. *Hydrocolloids in food processing*. Oxford: Willey-Blackwell (IFT Press).

Mantel, C. L. 1947. *The water soluble gums*. New York: Reinhold Publishing Corp.

Nishinari, K., and E. Doi. 1994. *Food hydrocolloids: Structure, properties and functions*. New York: Plenum Press.

Nussinovitch, A. 1997. *Hydrocolloid applications. Gum technology in the food and other industries*. London: Blackie Academic and Professional.

Nussinovitch, A. 2003. *Water soluble polymer applications in foods*. Oxford: Blackwell Publishing.

Nussinovitch, A. 2010. *Plant gum exudates of the world: Sources, distribution, properties, and applications*. Boca Raton: CRC Press, Taylor and Francis Group.

Nussinovitch, A., and M. Hirashima. 2014. *Cooking innovations, using hydrocolloids for thickening, gelling and emulsification*. Boca Raton, FL:CRC Press, Taylor & Francis Group.

Phillips, G. O., and P. A. Williams. 2009. *Handbook of hydrocolloids*. Cambridge: Woodhead Publishing Limited.

Smith, F., and R. Montgomery. 1959. *The chemistry of plant gums and mucilages*. New York: Reinhold Publishing Corp.

Stephen, A. M. 1995. *Food polysaccharides and their application*. New York: Marcel Dekker Inc.

The Holy Bible, New International Version. *Exodus* 16:13–5, 31; 25: 10.

Whistler, R. L., and J. N. BeMiller. 1993. *Industrial gums: Polysaccharides and their derivatives*, 3rd edn. San Diego: Academic Press.

Whistler, R. L., and C. LSmart. 1953. *Polysaccharide chemistry*. New York: Academic Press.

CHAPTER 2

Arabinoxylans

2.1 HISTORICAL BACKGROUND

Cereal arabinoxylans (AXs) were first described in 1927, by W. F. Hoffmann and R. A. Gortner, as a viscous gum present in wheat flour. Since then, food technologists and cereal chemists have learned much about them due to their scientific and industrial importance.

2.2 OCCURRENCE AND CONTENT

AXs are minor constituents of whole cereal grains, but are significant components of plant cell walls. The thin walls surrounding the cells in the starchy endosperm and aleurone layer of most cereals consist mainly of AXs (60%–70%), with the exception of endosperm cell walls of barley (20%) and rice (40%). Non-endospermic tissues of wheat, particularly the pericarp and testa, also have very high AX contents (64%). They are also found in bran and husk. AXs are the major non-starchy polysaccharides in cereal grains.

AXs are present in the most important cereal grains, i.e., barley, maize, millet, oat, rice, rye, sorghum, and wheat. The total AX content in grains varies from ~1.6% to 3.1% in wheat flour to ~30% in corn bran. A positive relationship between the amount of AX accumulated in wheat and the drying conditions has been reported. In addition, AX content may differ based on genetic and other environmental factors. AXs have been additionally described in other sources, such as bamboo shoots, banana peels, Ispaghula seed husk, linseed mucilage, pangola grass, psyllium, and rye grass. Since AX is the main dietary fiber constituent in the cereal grains that make up a large part of our diet, an awareness of its physiological effects is vital.

2.3 STRUCTURE

AX is a hemicellulose with a xylose backbone and arabinose side chains. It is a major fiber component in many cereal grains. AXs obtained from rice, sorghum, finger millet, and maize bran are more complex than those obtained from common cereals. They may enclose substantial amounts of 2-O-linked glucuronic acid residues and are referred to as glucurono(arabino)xylans. The molecular structure of the physiologically active polysaccharides of psyllium husk (*Plantago* species) is somewhat exceptional. The major fraction produced by aqueous extraction followed by acidification of the extract yields gel-forming polysaccharides that consist of neutral AXs containing 22.6% arabinose and 74.6% xylose residues, with only traces of other sugars. Further information on general molecular features, structural heterogeneity, and analysis and detection can be located elsewhere.

2.4 SOURCES AND MANUFACTURE

A frequent approach to isolating AXs involves aqueous extraction of these polymers from vegetative materials. When water is used as the extraction medium, the resultant extract includes water-soluble AX, together with water-soluble proteins, sugars, additional polysaccharides and minerals, which need to be eliminated through purification steps. Depending on the origin of the AXs, a substantial proportion of these polymers may not be water-extractable. In the intact cell wall, cross-linked AXs form a structural network that is not soluble in an aqueous environment. AX chains can also be covalently cross-linked to each other or to other cell-wall constituents; in these cases, AXs can be extracted from the cell walls with alkali solutions. Once isolated from the cell-wall matrix, AX is water-soluble.

Although extracting AX from wheat bran is complicated, it can be manufactured from wheat endosperm during the commercial processing of wheat flour. When wheat flour is processed to manufacture starch and gluten, the fiber component, which is mainly AX, is left in the byproduct, from which the AX-rich fiber can be extracted. Aqueous or alkali treatments are regularly used to extract AX from specific plant tissues. As stated, within the cell-wall matrix, AXs are covalently or noncovalently associated to each other or to other cell wall components. Different techniques have been described for AX separation from a variety of vegetative tissues. In general, extraction involves the inactivation of endogenous enzymes in an aqueous environment and the use of hydrolytic enzymes to remove proteins and starch from the extract. To date, attempts have been centered on the extraction of AXs from inexpensive agricultural byproducts of food manufacturing, for instance banana peels, cereal bran, corn cobs, and the fleshy tissue of sugar beet, where AXs are among the chief constituents. Enzymatic, chemical, and physical processes have been used to extract AX from rye, bran, and wheat, while AX-enriched fiber is obtained from wastewater produced in a wheat starch plant, and a feruloylated AX from the wastewater of maize nixtamalization. The heteroxylan fraction of maize bran can be extracted with acidic or alkaline solutions to generate water-soluble maize bran gum. AX-enriched, water-soluble maize bran gum has also been extracted from maize bran under mild alkaline conditions.

2.5 PHYSICOCHEMICAL PROPERTIES

The molecular weight values of AXs show sizeable variations, as they are dependent on extraction procedures, preparation purity, and methods of determination. The weight-average molecular weight (M_w) of water-extractable wheat AX was found to be 300 kDa, and this determination was done by the process of high-performance size-exclusion chromatography using a light-scattering detector. A M_w range of 220–700 kDa was also determined by sequential fractionation of polymers that are isolated from wheat flour with water. Alkali-extractable wheat endosperm was reported to have a higher M_w of 850 kDa, even though the extraction method might degrade the polymer chain. The solubility of AXs in aqueous solution is closely related to the presence of arabinose substituents along the xylan backbone, their pattern of distribution, and the chain length. AX molecules may have extended lengths, and in solution, they behave like locally stiff, semiflexible random coils. Due to their molecular weight and locally stiff semiflexible coil conformation, AX chains exhibit viscosity-building properties in solution. At low shear rates, AX solutions act like Newtonian fluids, whereas at high shear rates, they demonstrate shear-thinning features like that of pseudoplastic materials. Cereal AXs containing ferulic acid residues can form gel networks, which are stabilized by covalent cross-links between the feruloyl groups of neighboring chains. Strong gels are obtained with AXs having a high content of ferulic acid residues, a high molecular weight, and a high degree of substitution. Other factors such as concentrations of polymer and oxidizing agents, and the extent of hydrogen bonding also contribute to AX gel strength.

2.6 FUNCTIONAL CHARACTERISTICS

AX-rich fiber behaves like guar gum. Both are predominantly fermented in the cecum. Consequently, AX-rich fiber appears to perform like a fast-fermenting soluble fiber. The beneficial effects of soluble fiber on carbohydrate metabolism have been well documented. Guar gum was shown to improve the postprandial glucose response in people who were in good physical shape as well as in those with type II diabetes. In the latter, soluble fiber was shown to improve long-term glycemic control. AX-rich fiber extracted from the byproducts of wheat flour processing has valuable effects on metabolic control in people with type II diabetes and, importantly, the products enriched with AX fiber are edible. AX-rich fiber has the potential for inclusion in a broad range of food products that can be used for the dietary management of type II diabetes. Other soluble fibers, such as β-glucan (Chapter 4), pectin, psyllium, and *konjac-mannan*, have also been shown to be beneficial. A study in humans demonstrated that dietary addition of 10 g AX isolated from maize for 6 months improves glucose tolerance and hemoglobin A_{1c} concentrations in obese people with diabetes. Hemoglobin A_{1c} is a minor component of hemoglobin which binds glucose.

2.7 APPLICATIONS

AXs exhibit more than a few attractive practical properties, including viscosity augmentation, gel structuring, emulsion and froth stabilization, water incorporation, fat substitution, and prebiotic activity. They have an effect on the use of cereal grains in milling, brewing, bread production, and animal feed manufacture. As dietary fiber ingredients, AXs have an influence on the dietary quality of cereal foods; they present the nutritional advantages of both soluble and insoluble fibers, and, due to the presence of phenolic moieties in their molecular structure, they might also have a number of antioxidant properties.

Chronic diseases and obesity are on the rise. There is an established link between the consumption of dietary fiber and a variety of physical benefits, and this has raised consumer enthusiasm for foods supplemented with dietary fiber. AXs have dietary significance as a fiber constituent. Studies have revealed the beneficial effects of rye AX on cecal fermentation, creation of short-chain fatty acids, decrease in serum cholesterol, and improved adsorption of calcium and magnesium.

Feruloylated oligosaccharide from wheat flour demonstrated high antioxidant activity, antiradical effectiveness, and inhibition of copper-mediated oxidation of human low-density lipoprotein. AXs are not digested in the small intestine, but supply fermentable carbon resources for bacteria that inhabit the large bowel. Pretreatment of non-water-extractable wheat AX by xylanase supplied oligomers that are well exploited by the gut bacteria, thus extending the prebiotic index. Modified AX from rice bran was reported to be an effective inducer of macrophage phagocytic activity. Anti-HIV activity in vitro was also reported for this molecule.

Many food manufacturers have established ways to include AX in their manufactured goods. Foam is formed by trapping gaseous bubbles in a liquid or solid. Foam is normally an extremely complex system consisting of polydisperse gas bubbles separated by draining films. AX stabilizes foams by increasing both the viscosity and the elasticity of the film surrounding the gas cells. An emulsion is a mixture of two or more liquids which are normally immiscible (cannot be mixed). Emulsions belong to a more general class of biphasic systems called colloids. In an emulsion, one liquid (the dispersed phase) is dispersed in the other (the continuous phase). Examples of emulsions include vinaigrette, milk, and cutting fluid for metal working. AX also stabilizes emulsions by increasing the viscosity of the continuous phase. Such an increase in emulsion stability was reported for an oregano oil-in-water emulsion, when the AX concentration was increased from 0% to 1% (w/v). AXs are also used to form films due to their capacity to create a consistent matrix. The mechanical and barrier properties of AX films are somewhat similar to those of other films produced from gluten, whey protein isolates, hydroxypropyl methylcellulose, methylcellulose, or starch.

For the last 50 years, the use of AX in bread-making has been a hotly debated practice. AXs are typically divided into water-extractable and non-water-extractable categories. Water-extractable AXs also affect gas retention by the dough. Carbon dioxide manufactured within the dough by a leavening agent dissolves in the aqueous phase and then diffuses into gas cells formed by the gluten structure. Water-extractable AXs combine with the gluten proteins to form a macromolecular complex that slows down carbon dioxide diffusion out of the dough, which results in the collapse of proof volume. On the other hand, non-water-extractable AXs damage dough quality and the value of baked goods. Owing to their high water-absorption properties, these AXs swell and produce bulky agglomerates, interrupting the continuity of the complex gluten films surrounding the gas cells.

In the last decade, AXs have once again become the focus of much attention owing to the growing use of endoxylanases in European wheat and rye flour bread-making processes. The renewed awareness has led to substantial advances in our understanding of both AX and endoxylanase functionalities in bread-making. Permanent gels can be formed from an aqueous extract of wheat flour AX by treating it with oxidizing agents at an ambient temperature. It was shown that adding xylanase to the formula of a barley-enriched bread improved bread quality and soluble fiber content. Recently, a method was patented for increasing the level of water-soluble AX, which has an average degree of polymerization in the range of 5–25, in a baked product. The method includes several steps: preparing the dough with flour or a mixture of flours, with milling fractions having a total AX content of at least 2.0%, and adding an enzyme preparation to the dough that includes at least one thermophilic endoxylanase. An advantage to this method is that baked products can be obtained with at least 1.7% of the desired type of AX. The baked products obtained from this method have been shown to have health benefits.

Polysaccharides such as gums, oat β-glucan (Chapter 4), and AX have been studied for their physicochemical properties and have been extensively used as improvers in bread-making. Addition of certain polysaccharides as dietary fiber to bread ingredients will usually enhance the bread's nutritive value, but they may also reduce the quality of the final product by conferring excessive viscosity or stickiness. In particular, they reduce the storage stability of bread. On the other hand, as celiac disease—a wheat allergy—becomes more prevalent, utilization of these polysaccharides in place of wheat flour to make a gluten-like dough is becoming an increasingly important practice in bread-making.

Dough is a paste prepared from any cereal (grain) or leguminous crop by mixing its flour with a small amount of water and/or other liquid. This process is a precursor to making a vast variety of foodstuffs, namely breads and bread-based items, including flatbreads, noodles, crusts, pastry, and the like. Refrigerated dough quality is very important for storage. Under refrigeration, fluid can separate from the dough and form a syrup-like liquid that can leak from the package. This is termed as "dough syruping" and is not tolerated by the consumer. The problem might be connected to the AXs and endoxylanases in the flour. The dark-yellow fluid derives from the usual components of flour and includes water and water-soluble constituents of the dough. The loss of water implies that the dough's water absorption and retention abilities decrease during storage. Syrup formation in refrigerated dough has been reported to be due to the degradation of high-molecular-weight AXs and the conversion of non-water-extractable AXs to water-extractable ones, which have relatively lower water-holding capacity. Addition of xanthan gum significantly decreased the amount of water released by the AX molecules in refrigerated dough. Dough consistency was most stable at 0.5% xanthan gum in the dough formulation. The xanthan gum lowered the water activity of the food as it competed for water with the other biopolymers in the formulation, such as protein and starch.

Beer is the world's most widely consumed, and probably the oldest, alcoholic beverage; it is the third most popular drink overall, after water and tea. It is produced by the brewing and fermentation of sugars mainly derived from malted cereal grains, most commonly malted barley and malted wheat. In the beer industry, AXs are added for viscosity and membrane filterability. Problems are more pronounced when wheat and wheat malt are used as adjuncts due to their higher AX contents and higher molecular weights.

AXs produce very highly viscous solutions and gels. In addition to covalent cross-linking, non-covalent bonds between AX chains may be formed as well, which add to the AX gel structure. AX gels have potential applications in colon-specific protein delivery for three main reasons: their macroporous structure with 200–400-nm mesh size, their aqueous surroundings, and their dietary fiber character. Bovine serum albumin entrapped within AX gels can be protected from in vitro pepsin proteolysis. AX gels can also load and release proteins with molecular masses ranging from 43 to 669 kDa. When 0.1% insulin was entrapped, a low quantity of it was released by diffusion. Most of the protein was liberated only after gel degradation by colonic bacteria. Thus, AX gels can be suitable matrices for insulin delivery to specific sites, such as the colon.

Subjecting the external shell of rice bran to enzymes from an extract of *Hyphomycetes* mushroom mycelia produces an AX compound. This compound, called MGN-3 (or BioBran® in Japan), is a complex with AX as its main constituent. The Biobran MGN-3 AX compound enhances immune responses in diabetic and cancer patients. MGN-3 may also be important in treating AIDS patients or those undergoing chemotherapy. Even though presented by the manufacturers as a generally safe substance devoid of side effects at recommended doses, the United States Food and Drug Administration (FDA) issued a court order in

2004 against the marketing of MGN-3, which remains valid today, charging that it had been promoted as a drug for the treatment of cancer, diabetes, and HIV.

2.8 RECIPES

2.8.1 Soda Bread (Figure 2.1)

(Makes 1 loaf)

450 g (3 ¼ cups) all-purpose flour
50 g AX powder (10% AX)
8 g (2 tsp) baking soda
6 g (1 tsp) salt
300–400 g (1⅕–1⅔ cups) buttermilk (Hint 2.8.1.1)

1. Sift flour, AX powder, and baking soda into a bowl (Figure 2.1A), and then mix in salt.

Figure 2.1 Soda bread. (A) Sift flour, AX powder, and baking soda into a bowl, and mix in salt. (B) Make a well in the center of the flour mixture and pour buttermilk. (C) Stir in buttermilk. (D) Form dough and put in a bread pan. (E) Mark a deep cross in the dough with a knife. (F) Cool after baking.

2. Make a well in the center of the flour mixture (1), pour buttermilk into it (Figure 2.1B), and stir with a spatula until dough becomes sticky (Figure 2.1C) (Hint 2.8.1.2).
3. Place the dough (2) on a lightly floured cutting board, and knead it lightly for a few minutes (Hint 2.8.1.3).
4. Preheat the oven to 200°C. Line a bread pan with baking paper, and sprinkle lightly with flour.
5. Shape the dough (3) into a ball (Figure 2.1D), put it in the bread pan, and make two slits in the shape of a cross with a knife (Figure 2.1E).
6. Bake the dough (5) at 200°C for 40–45 minutes.
7. Cool the bread on a cake cooler or wire rack (Figure 2.1F).

pH = 5.94

Preparation hints:

2.8.1.1 Buttermilk is usually preferred for soda bread, but you can also use milk, live yoghurt, or soured milk as a substitute for the buttermilk.

2.8.1.2 If the dough is too soft and sticky add all-purpose flour, and if it is dry and floury add buttermilk.

2.8.1.3 Do not over-knead. Soda bread dough is not smooth and elastic like regular bread dough.

This is a recipe for traditional non-leavened (non-yeast) Irish bread. It is eaten as an accompaniment at breakfast.

2.8.2 Cake Muffins (Figure 2.2)

(Makes 5 muffins)

100 g (a little under ¾ cup) all-purpose flour

4 g (1 tsp) baking powder

20 g AX powder (10% AX)

50 g unsalted butter, softened

60 g (¼ cups) granulated sugar

1 g (¼ tsp) salt

1 egg
55 mL (a little over a ¼ cup) milk
Vanilla oil or vanilla flavoring, butter for greasing the pan

1. Allow egg and milk to come to room temperature.
2. Sieve flour, AX powder, and baking powder into a bowl (Figure 2.2A).
3. In a separate bowl, whisk the softened butter until it becomes creamy (Figure 2.2B), gradually add sugar and stir well until the mixture becomes whiter and softer (Figure 2.2C).
4. Mix salt and vanilla into the butter and sugar mixture (3).
5. Beat egg and pour it into the mixture (4) in several steps (Hint 2.8.2.1), and stir well with a whisk (Figure 2.2D).
6. Preheat oven to 180°C and lightly grease muffin pan or use disposable muffin cups.

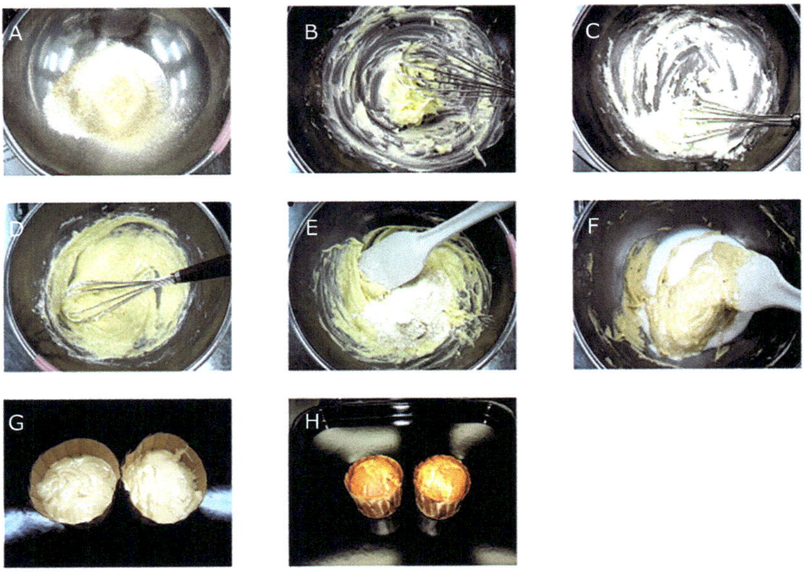

Figure 2.2 Cake muffins. (A) Sift flour, AX powder, and baking powder into a bowl. (B) Whisk softened butter in another bowl. (C) Add sugar to the butter and whisk. (D) Add beaten egg. (E) Add the flour mixture to the creamed butter, sugar, and egg. (F) Stir in milk. (G) Pour batter into muffin cups. (H) Bake.

7. Add 1/3 of the sieved flour (2) to the creamed mixture (5) (Figure 2.2E) and stir. Mix in 1/3 of the milk (Figure 2.2F), and then add the remaining flour and milk alternately in two steps, mixing after each step.
8. Pour the batter (7) into the muffin cups (Figure 2.2G), and bake at 180°C for 20–25 minutes (Figure 2.2H) (Hint 2.8.2.2).
9. Cool the muffins (8) on a cake cooler or wire rack.

pH of batter = 5.33

Preparation hints:

2.8.2.1 Stir well after each addition, aiming for an airy batter.

2.8.2.2 To check if it is ready, poke the center of a muffin with a toothpick or skewer; if it comes out clean, the muffins are done.

This recipe makes an American-style muffin, which is similar to a cupcake. You can add your own favorite ingredients, such as raisins, dried fruit with rum, chocolate, sweet potato, bacon, and onion.

2.8.3 Chocolate Cookies (Figure 2.3)

(Makes 20–30 cookies)

100 g (a little under ¾ cup) all-purpose flour
1.5 g (⅓ tsp) baking powder
20 g AX powder (10% AX)
50 g butter, softened
20 g (1 ½ tbsp) granulated sugar
1 g (⅙ tsp) salt
30 mL (2 tbsp) milk
Chocolate for topping

1. Whisk softened butter to a cream-like consistency, and then mix in sugar.
2. In a separate bowl, sieve flour, AX powder, and baking powder.
3. In a third bowl, dissolve salt in milk.

FIGURE 2.3 CHOCOLATE COOKIES. (A) STIR SALTED MILK INTO A SIEVED FLOUR MIXTURE. (B) STIR IN THE BUTTER AND SUGAR MIX. (C) WRAP THE DOUGH AND LEAVE TO CHILL IN THE REFRIGERATOR. (D) ROLL OUT DOUGH. (E) CUT OUT COOKIES. (F) POKE HOLES IN THE TOP OF THE COOKIE WITH A FORK. (G) BAKE. (H) MELT CHOCOLATE IN A BOWL OVER HOT WATER. (I) SPREAD MELTED CHOCOLATE OVER COOKIES.

4. Pour salted milk (3) into the sieved flour mixture (2) in several steps (Figure 2.3A), stirring after each step.
5. Add the butter and sugar mixture (1) and stir with a spatula (Figure 2.3B).
6. Wrap the dough in a plastic cling film (Figure 2.3C), and refrigerate for at least 30 minutes.
7. Roll out the dough (6) to 3-mm thickness with a rolling pin (Figure 2.3D), form cookies with a cookie cutter (Figure 2.3E) (Hint 2.8.3.1), and poke holes in the top of the cookie with a fork (Figure 2.3F).
8. Preheat oven to 170°C, and bake cookies (7) for 15 minutes (Figure 2.3G).
9. Cool the cookies on a cake cooler or wire rack.

10. Melt chocolate over a hot water bath, slightly bigger than the bowl (Figure 2.3H), and spread over biscuits (9) (Figure 2.3I).

pH of batter = 5.00

Preparation hints:

2.8.3.1 This should be done quickly because warm dough becomes sticky and difficult to handle.

These cookies have a pleasant texture—crunchy and crumbly. For variety, use different chocolates or change the shape of the biscuits.

2.9 TIPS FOR THE AMATEUR COOK AND PROFESSIONAL CHEF

- AXs in wheat flour have disproportionately high water-absorption properties (22%) relative to their absolute content.
- When AX types are adjoined in wheat flour, both water-soluble and non-water-soluble AXs take up large quantities of water, thus depleting the accessible pool for appropriate gluten development and film formation. This problem can be corrected by adding more water, or by adding a xylanase-containing enzyme preparation, which causes the release of AX-bound water.
- Water-soluble AXs increase the viscosity of the dough's aqueous phase and have a positive effect on dough structure and stability.
- Loaf volume and crumb structure are enhanced by the addition of water-soluble AXs.
- Excess amounts of AX may cause viscosity build up and negate their beneficial effects.
- Non-water-extractable AXs depress loaf volume and create coarser and firmer crumbs. Such negative effects can be reversed by using endoxylanases with specificity activity toward the water-insoluble AXs.
- Optimal doses of xylanases improve dough consistency, fermentation stability, over-rising, loaf volume, and crumb structure and softness.

References and Further Reading

Amodo, R., and H. Neukom. 1985. Minor constituents of wheat flour: the pentosans. In *New approaches to research on cereal carbohydrates*, ed. R. D. Hill and L. Munch, 241–51. Amsterdam: Elsevier Science Publishers.

Baker, J. C., Parker, H. K., and M. D. Mize. 1943. The pentosans of wheat flour. *Cereal Chem.* 20:267–80.

Barr, D. J. 1989. The engineering of a modern wheat starch process. In *Wheat is unique: Structure, composition, processing, end-use properties, and products*, ed. Y. Pomeranz, 501–5. St Paul, MN: American Association of Cereal Chemists.

Berlanga-Reyes, C. M., Carvajal-Millán, E., Lizardi-Mendoza, J., Rascón-Chu, A., Marquez-Escalante, J. A., and A. L. Martínez-Lòpez. 2009. Maize arabinoxylan gels as protein delivery matrices. *Molecules* 14:1475–82.

Biliaderis, C. G., Izydorczyk, M. S., and O. Rattan. 1995. Effect of arabinoxylans on bread-making quality of wheat flours. *Food Chem.* 53:165–71.

Chanliaud, E. 1995. Extraction, caractérisation et propriétés fonctionnelles des hétéroxylanes de son de Maïs. Dissertation, University of Paris VII and ENSIA.

Cordenunsi, B. R., Misuzu, S. T., and F. Lajolo. 2008. Non-starch polysaccharide composition of two cultivars of banana (*Musa acuminata* L.: cvs. Mysore and Nanicaõ). *Carbohydr. Polym.* 71:26–31.

Durham, R. K. 1925. Effect of hydrogen peroxide on relative viscosity measurements of wheat and flour suspensions. *Cereal Chem.* 2:297–305.

Fairchild, R. M., Ellis, P. R., Byrne, A. J., Luzio, S. D., and M. A. Mir. 1996. A new breakfast cereal containing guar gum reduces postprandial plasma glucose and insulin concentrations in normal-weight human subjects. *Br. J. Nutr.* 76:63–73.

Faurot, A. L., Saulnier, L., Bérot, S., Popineau, Y., Petit, M. D., Rounau, X., and J. F. Thibault. 1995. Large scale isolation of water-soluble and water-insoluble pentosans from wheat flour. *Leb. Wiss. Technol.* 28:436–44.

Fincher, G. B., and B. A. Stone. 1974. A water-soluble arabinogalactan peptide from wheat endosperm. *Aust. J. Biol. Sci.* 27:117–32.

Ford, C. W. 1989. A feruloylated arabinoxylan liberated from cell walls of *Digitaria decumbens* (pangola grass) by treatment with borohydride. *Carbohydr. Res.* 190:137–44.

Fuessl, H. S., Williams, G., Adrian, T. E., and S. R. Bloom. 1987. Guar sprinkled on food: effect on glycaemic control, plasma lipids and gut hormones in non-insulin dependent diabetic patients. *Diabet. Med.* 4:463–8.

Groop, P. H., Aro, A., Stenman, S., andL. Groop. 1993. Long-term effects of guar gum in subjects with non-insulin-dependent diabetes mellitus. *Am. J. Clin. Nutr.* 158:513–8.

Hanai, H., Ikuma, M., Sato, Y, Iida, T., Hosoda, Y., Matsushima, I., Nogaki, A., Yamada, M., and E. Kaneko. 1997. Long-term effects of water-soluble corn bran hemicellulose on glucose tolerance in obese and nonobese patients: Improved insulin sensitivity and glucose metabolism in obese subjects. *Biosci. Biotechnol. Biochem.* 61:1358–61.

Hartley, R. D., and E. C. Jones. 1976. Diferulic acid as a component of cell walls of *Lolium multiflomm. Phytochemsitry.* 15:1157–60.

Henry, R. J. 1985. A comparison of the non-starch carbohydrates in cereal grains. *J. Sci. Food Agric.* 36:1243–53.

Hopkins, M. J., Englyst, H. N., Macfarlane, S., Furrie, E., Macfarlane, G. T., and A. J. McBain. 2003. Degradation of cross-linked arabinoxylans by the intestinal microbiota in children. *Appl. Environ. Microbiol.* 69:6354–60.

Izydorczyk, M. S., and C. G. Biliaderis. 1995. Cereal arabinoxylans: Advances in structure and physicochemical properties. *Carbohydr. Polym.* 28:33–48.

Jenkins, D. J., Goff, D. V., Leeds, A. R., Alberti, K. G., Wolever, T. M. S, Gassull, M. A., and T. D. Hokaday. 1976. Unabsorbable carbohydrates and diabetes: decreased post-prandial hyperglycaemia. *Lancet.* 2:172–4.

Jenkins, D. J., Leeds, A. R., Gassull, M. A., Cochet, B., andG. M. Alberti. 1977. Decrease in postprandial insulin and glucose concentrations by guar and pectin. *Ann. Intern. Med.* 86:20–3.

Jenkins, D. J. A., Wolever, T. M. S., Leeds, A. R., Gassull, M. A., Haisman, P., Dilawari, J., Goff, D. V., Metz, G. L., and K. G. Alberti. 1978. Dietary fibre, fibre analogues, and glucose tolerance: Importance of viscosity. *Br. Med. J.* 1:1392–4.

Kweon, M., Slade, L., and H. Levine. 2009. Oxidative gelation of solvent-accessible arabinoxylans in the predominant consequence of extensive chlorination of soft wheat flour. *Cereal Chem.* 86:421–4.

Landin, K., Holm, G., Tengborn, L., and U. Smith. 1992. Guar gum improves insulin sensitivity, blood lipids, blood pressure, and fibrinolysis in healthy men. *Am. J. Clin. Nutr.* 56:1061–5.

Lu, Z. X., Walker, K. Z., Muir, J. G., and K. O'Dea. 2004. Arabinoxylan fibre improves metabolic control in people with Type II diabetes. *Eur. J. Clin. Nutr.* 58:621–8.

Maires, D. J., and B. A. Stone. 1973. Studies on wheat endosperm. I. Chemical composition and ultrastructure of the cell walls. *Aust. J. Biol. Sci.* 26:793–812.

Nino-Medina, G., Carvajal-Millan, E., Rascon-Chu, A., Marquez-Escalante, J. A., Guerrero, V., and E. Salas-Munoz. 2010. Feruloylated arabinoxylans and arabinoxylan gels: structure, sources and applications. *Phytochem. Rev.* 9:111–20.

Nussinovitch, A. 1997. *Hydrocolloid applications. Gum technology in the food and other industries.* London: Blackie Academic & Professional.

Nussinovitch, A. 2003. *Water soluble polymer applications in foods.* Oxford: Blackwell Publishing.

Nussinovitch, A. 2010. *Plant gum exudates of the world sources, distribution, properties and applications.* Boca Raton, FL: CRC Press, Taylor & Francis Group.

Nussinovitch, A., and M. Hirashima. 2014. *Cooking innovations, using hydrocolloids for thickening, gelling and emulsification.* Boca Raton, FL: CRC Press, Taylor & Francis Group.

Ring, S. R., and R. R. Selvendran. 1980. Isolation and analysis of cell wall material from beeswing wheat bran (*Triticum aestivum*). *Phytochemistry.* 19:1723–30.

Saghir, S., Iqbal, M. S., Hussain. M. A., Koschella, A., and T. Heinze. 2008. Structure characterization and carboxymethylation of arabinoxylan isolated from Ispaghula (*Plantago ovata*) seed husk. *Carbohydr. Polym.* 74:309–17.

Simsek, S. 2009. Application of xanthan gum for reducing syruping in refrigerated doughs. *Food Hydrocolloids.* 23:2354–8.

Sinha, A. K., Kumar, V., Makkar, H. P. S., De Boeck, G., and K. Becker. 2011. Non-starch polysaccharides and their role in fish nutrition—A review. *Food Chem.* 127:1409–26.

Slavin, J. 2007. Dietary carbohydrates and risk of cancer. In *Functional food carbohydrates*, 1st edn., ed. M. S. Izydorczyk and C. G. Biliaderis, 371–85. Boca Raton: CRC Press.

Southgate, D. A. T., and H. Englyst. 1985. Dietary fiber: Chemistry, physical properties and analysis. In *Dietary fiber, fiber depleted foods and disease*, ed. H. Trowell, D. Burkitt and K. Heaton,31–55. London: Academic Press Inc.

Van Haesendonck, I. P. H. V., Broekaert, W. F., Georis, J., Delcour, J., Courtin, C., and F. Arnaut. 2010. Bread with increased arabinoxylo-oligosaccharide content. U.S. Patent No. 20,100,040,736.

Vuksan, V., Jenkins, D. J., Spadafora, P., Sievenpiper, J. L., Owen, R., Vidgen, E., Brighenti, F., Josse, R., Leiter, L. A., and C. Bruce-Thompson. 1999. Konjac-mannan (glucomannan) improves glycemia and other associated risk factors for coronary heart disease in type 2 diabetes. A randomized controlled metabolic trial. *Diabetes Care* 22:913–9.

Whistler, R. L. 1993. Hemicelluloses. In *Industrial gums, polysaccharides and their derivatives*, 1st edn., ed. R. L. Whistler and J. N. BeMiller. Orlando: Academic Press.

Wolever, T. M. S., and D. J. A. Jenkins. 1993. Effect of dietary fiber and foods on carbohydrate metabolism. In: *CRC handbook of dietary fiber in human nutrition*, 2nd edn., ed. G. A. Spiller, 111–52. Boca Raton: CRC Press.

Wolever, T. M., Vuksan, V., Eshuis, H., Spadafora, P., Peterson, R. D., Chao, E. S., Storey, M. L., and D. J. Jenkins. 1991. Effect of method of administration of psyllium on glycemic response and carbohydrate digestibility. *J. Am. Coll. Nutr.* 10:364–71.

Wursch, P., and F. X. Pi-Sunyer. 1997. The role of viscous soluble fiber in the metabolic control of diabetes. A review with special emphasis on cereals rich in beta-glucan. *Diabetes Care* 20:1774–80.

CHAPTER 3

Bacterial Cellulose

3.1 INTRODUCTION

Cellulose, a constituent of the plant skeleton, is produced by land plants. It is also a major component of the cell walls of practically all the plants, fungi, and various algae. Cellulose, which has been pulverized by either chemical or physical treatments, can be utilized as a food additive. Plant-derived cellulose contains both hemicellulose and lignin, but cellulose produced by bacteria, also termed as bacterial cellulose (BC), is cellulose in its pure form. Cellulose can also be produced by microorganisms such as *Acetobacter*, which is capable of manufacturing vinegar. Other producers of cellulose belong to the genera *Rhizobium*, *Agrobacterium*, and *Sarcina*. Upon production of vinegar, a gel-like membrane often forms on the surface of the culture fluids, composed of BC. For many years, BC has been consumed as a dessert called "*nata de coco*".

3.2 MANUFACTURE AND REGULATORY STATUS

Non-pathogenic bacteria, such as the subspecies of *Acetobacter aceti*, produce BC by fermentation, and this BC is different from the fibrillated or crystallized cellulose manufactured from the pulp. The fermentation proceeds in five stages: (1) a culture of the model bacterium *Acetobacter xylinum* is added to sterile growth medium at pH 5.0 and at a temperature of 30°C; (2) one volume from the first stage is added to 20 volumes of fresh sterilized medium, and the bacterium is grown to an optical density (OD) > 1; (3) when the culture reaches OD > 1.5, the medium is moved to four preseeded flasks, serving as the primary seed fermentor. These are maintained at 30°C, under conditions of sufficient oxygen; (4) the fermentation from stage 3 is used to inoculate the secondary seed fermentor, with a volume of ~4,000 L. Under strict conditions of pH, temperature, and dissolved oxygen, cellulose strands become visible; (5) this secondary seed is used to inoculate the final fermentor, having a volume that is fivefold greater. In addition to the aforementioned conditions, the amount of carbon dioxide generated, the glucose and cellulose contents, and the cell growth are monitored. After 59 hours, the glucose supply is turned off and after one additional hour, fermentation is stopped and harvesting begins. The purification and recovery starts with dewatering of the fermentor broth by a twin wire belt press, until a cake of ~20% solids is attained. The cake is reslurried to ~1.5%–3.0% solids. Solid NaOH is added to raise the pH to 13.1, and the slurry is heated to 65°C for 2 hours with agitation to dissolve the microorganisms. After an additional 2 hours, the pH is lowered with sulfuric acid to 6.0–8.0. To purify the BC, the slurry is passed through two more cycles of reslurrying and dewatering. BC is a generally recognized as safe (GRAS) food ingredient (GRAS affirmation petition filed on December 11, 1991, and accepted by the FDA on April 13, 1992).

3.3 STRUCTURE

The chemical structure of BC is the same as that of cellulose from plants. It is a straight-chain polysaccharide whose D-glucose units are connected by β-1,4-bonding. BC consists of an ultrafine network of cellulose nanofibers (3–8 nm) which are highly uniaxially oriented. This three-dimensional (3D) structure is highly crystalline (60%–80%) and confers remarkable mechanical and physicochemical properties. It is also essential to note that the fiber diameter of BC is approximately one-hundredth that of plant cellulose, while its Young's modulus is approximately the same as that of aluminum.

3.4 TECHNICAL DATA

Although BC is not soluble in water, it can be dispersed in water to create a network. After swelling, a continuous 3D-network structure of partly connected cellulose fibers is formed. This 3D structure holds water and develops viscosity. In comparison, cellulose from plants is similarly insoluble in water but exists in the form of fine particles, which are poor at forming a 3D structure that can hold water. Thus the plant cellulose never becomes viscous and instead precipitates, except when combined with a polysaccharide. After forming a 3D-network structure, BC solutions behave as pseudoplastic (shear-thinning) fluids. BC is weakly reactive with various additives since it is not water soluble. The 3D-network structure is little affected by heat, acids, or salts. Furthermore, the network provides good dispersion stability for insoluble substances. In addition to the pseudoplastic viscosity demonstrated by liquids that contain swollen BC, these fluids also show good suspension and dispersion-stability capabilities, as a result of the strong 3D structure. The viscosity of fluids that contain swollen BC depends only weakly upon temperature. To swell BC powder in water, a strong shear force, without heating, is required. Swelling can also be achieved by a common propeller mixer and by homogenization. After swelling, salt addition or pH change has no effect on the fluid properties and, therefore, addition of ingredients, pH adjustment, heat treatment, or retorting should be performed only after the formation of a fully established and stable 3D network.

3.5 USES AND APPLICATIONS

The inclusion of very small amounts of BC in food may provide good dispersion and emulsion stability. Furthermore, such foods might acquire a good mouthfeel, and this is dependent upon good shape retention. These qualities result from the formation of the 3D fiber structures that remain stable in a medium containing acids and/or salts and when treated with heat. BC is used to stabilize dispersions, suspensions, and emulsions as a fat-replacing agent (i.e., calorie reducer), as a thickener and texturizer, and to prevent protein aggregation, among many other functions. The use of BC as a cryoprotectant and a support for the immobilization of probiotic lactic acid bacteria is a new concept with great potential. BC in the form of *nata* is nearly indigestible and therein lies its importance. BC can be employed in other fermentations, such as acetic acid fermentation, and it can also be used as an inert support in the second step of vinegar production, i.e., conversion of ethanol to acetic acid.

Adding BC to beverages, such as cocoa drinks (as a stabilizer for insoluble solid matter), powdered green tea (to improve heat stability) and the like, can prevent precipitation. This is because its insoluble network structure of very fine cellulose fibers takes up the insoluble solid matter, preventing it from completely precipitating out. Due to the yield stress of the fluid, it also has very good dispersive ability. In addition, inclusion of a small amount of BC has only a small influence on viscosity and, thus, will not provide a thick or distinct mouthfeel due to its insolubility.

BC can also be used in calcium-fortified drinks (to achieve a good mouthfeel without raising viscosity), in non-oil dressings (as a stabilizer for insoluble matter, to achieve a desired flowability, to improve stability, and/or to achieve a fatty mouthfeel), or in sauces for roasted meat (as a stabilizer to decrease threading, to achieve a non-gluey, pleasant mouthfeel, to improve stability and stickiness). Adding BC (*nata*) to Chinese-style meatballs resulted in greater cooking loss and softening effects on texture. These adverse effects were more pronounced when 20% or more BC was used. However, addition of 10% *nata* to Chinese-style meatballs created a product that seemed similar to the control meatball, in respect to sensory and shelf stability properties with no adverse effect on texture. In addition, it conferred acceptable juiciness and chewiness to the meatballs, making *nata* a possible useful ingredient in Chinese-style emulsified meat products.

When BC is added to frozen desserts that include milk proteins, the 3D network that forms in the soft-mix takes up and stabilizes the milk protein. Other hydrocolloids can be included to minimize whey separation. Thus, in products, such as lacto-ice, BC enhances creamy mouthfeel and shape retention, and in soft-mix, its addition prevents separation of whey and improves mouthfeel. The consequences of incorporating a range of concentrations of BC on the physicochemical and sensorial properties of the low-fat, soft cheese, *Turkish Beyaz,* were studied during a 60-day ripening period. Fat reduction and the use of BC for cheese-making improved some properties of *Turkish Beyaz* cheese and prevented the cheese blocks from melting. Moisture, pH, and the instrumental hardness of reduced-fat cheeses increased with and without BC; however, BC decreased hardness and improved sensory properties (only 1.7% BC) in comparison with reduced-fat cheeses. Therefore, BC is useful as a fat mimetic as well as for enhancing cheese quality.

BC enables the retorting of chilled pudding with no protein aggregation. Its use together with a gelling agent in the pudding confers resistance to the retort process due to the 3D-network structure as explained above; this structure stabilizes the milk protein by taking it up into the network, and it physically suppresses milk protein aggregation by separating the protein molecules beyond the distance that allows them to aggregate. The 3D-network structure also helps disperse pieces of fruit (i.e., semisolid or solid particles) in jellies, even if the preparation involves heating to high temperatures. In dressings, BC stabilizes the seasonings and improves the mouthfeel more effectively than xanthan gum. BC can also achieve dispersion stability in sauces. Finally, in mayonnaise and dressings, BC can very effectively substitute starch as a fat replacement. Starch addition produces a characteristic odor, and starch does not have good meltability in the mouth, it is poor at releasing flavor, and it creates a thick mouthfeel. In contrast, BC supports such products with good body, and a very creamy and thin mouthfeel. It also imparts good meltability and improves flavor release; in addition, its physical properties are very slightly changed by heat and it, thus, confers stability to the products in which it is included.

3.6 RECIPES

3.6.1 Introduction

Although BC has important applications in a variety of food formulations, its use in the kitchen at this stage is minimal. Therefore, we first present a protocol for the commercial preparation of *nata de coco* (see Section 3.6.2), and then give recipes for its use as a dessert base (see Section 3.6.3). *Nata de coco* is a popular dessert in the Philippines, which is also exported to Japan. The food is produced by *Gluconacetobacter xylinus* using coconut milk as the carbon source. Another popular BC-containing food product, which can be prepared in the home kitchen, is the Chinese *kombucha* or Manchurian tea, obtained by growing yeast and *Gluconacetobacter* (*Acetobacter*) in a medium containing tea extract and sugar (see Section 3.6.4).

3.6.2 Commercial "*Nata de coco*" Preparation

(Makes 20–25 kg)

1 kg ripe coconut
2 L water
2 kg white sugar
500–600 g glacial acetic acid
6 L mother liquor
26 L water

1. Grate coconut, add 1 L of water, and squeeze the coconut meat to extract the milk.
2. Filter the extract to remove residue.
3. Add 1 L water to the coconut meat and extract the coconut milk again.

4. Mix the first (1) and second (3) extracts and dilute with 26L water.
5. Add white sugar, acetic acid, and mother liquor to the coconut milk (4) and mix well until sugar dissolves.
6. Pour the mixture (5) into containers with 2–3-cm thick walls.
7. Cover containers with paper and tighten with rubber bands.
8. Place the containers in a room at around 30°C for 8–10 days with occasional stirring.
9. When the surface of the mixture has set, you have *nata de coco*; when the *nata de coco* becomes 1.0–1.5-cm thick, harvest it.

pH of *nata de coco* = 3.53

This recipe is not prepared in the home kitchen. Manufactured *nata de coco* is sold as a canned or bottled food (Figure 3.1). Various *nata de coco* desserts are commercially available (Figure 3.2).

Figure 3.1 Manufactured *nata de coco* is sold as a canned food.

Figure 3.2 Various commercially sold *nata de coco* desserts.

3.6.3 *Nata de coco* Recipes

3.6.3.1 *Nata de coco* and Fruit (Figure 3.3)

(Serves 5)
200 g *nata de coco*
200 g canned fruit
100 g canned fruit syrup
60 g (⅓ cup) white sugar
100 mL (⅖ cup) water
10 mL (⅔ tsp) lemon juice (optional)
Mint leaves (optional)

1. Mix canned fruit syrup, white sugar, and water in a pan and heat until sugar dissolves (Figure 3.3A).
2. Remove from the heat, add lemon juice and/or mint flavoring if you like, and cool in an ice-water bath.
3. Cut canned fruit into bite-sized pieces, drain the *nata de coco*, and mix the two (Figure 3.3B).
4. Place some *nata de coco* and cut fruit (3) in a small bowl, pour some cold syrup over it (2), and garnish with mint leaves (Figure 3.3C).

Figure 3.3 *Nata de coco* and fruit. (A) Heat canned fruit syrup, white sugar, and water in a pan until the sugar dissolves. (B) Mix cut fruit and drained *nata de coco*. (C) Pour cold syrup over the dessert and garnish.

3.6.3.2 *Nata de coco* in Fruit Jelly (Figure 3.4)

(Serves 5)
200 g *nata de coco*
7.5–12 g gelatin powder

FIGURE 3.4 *NATA DE COCO* IN FRUIT JELLY. (A) DISSOLVE SOAKED GELATIN IN HEATED FRUIT JUICE. (B) POUR GELATIN DISPERSION INTO A BOWL AND COOL. (C) UPON THICKENING, GARNISH.

75 mL (5 tbsp) water
600 g (2 ⅖ cups) fruit juice
Mint leaves (optional)

1. Soak gelatin in water for 20 minutes.
2. Drain *nata de coco*.
3. Heat fruit juice at 80°C, remove from heat, add the soaked gelatin (1) (Figure 3.4A), and allow it to dissolve.
4. Pour the gelatin dispersion (3) into a bowl, and cool in an ice-water bath with stirring (Figure 3.4B).
5. When the gelatin dispersion (4) thickens, add the *nata de coco* (2), pour the mixture into glass bowls, and cool in the refrigerator or in an ice-water bath.
6. Garnish with mint leaves (Figure 3.4C).

3.6.3.3 MILK JELLY WITH *NATA DE COCO* (FIGURE 3.5)

(Serves 5)
2.5 g agar-agar powder
250 mL (1 cup) water
3.5 g gelatin powder
55 mL (3⅔ tbsp) water
55 mL (3⅔ tbsp) milk
30 g (3⅓ tbsp) white sugar
50 g blueberry jam*

* Any flavor of jam can be used.

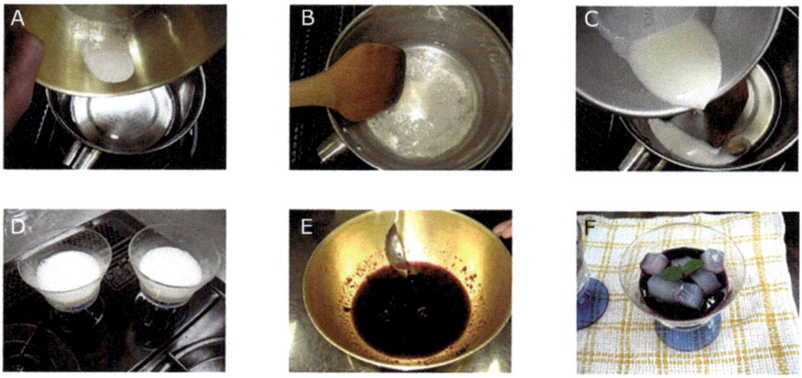

Figure 3.5 Milk jelly with *nata de coco*. (A) Disperse agar in water. (B) Dissolve agar while heating. (C) Combine the milk and vanilla extract with the agar-gelatin solution. (D) Pour into cups and cool. (E) Mix blueberry jam and Kirshwasser. (F) Pour fruit sauce over set jelly and garnish.

15 g (1 tbsp) Kirshwasser or *nata de coco* syrup*
120 g *nata de coco*
Vanilla extract, mint leaves

1. Soak gelatin in 55 mL water for 20 minutes.
2. Add agar-agar to 250 mL water (Figure 3.5A), heat to dissolution, and continue heating for 1 minute on medium heat (Figure 3.5B).
3. Add white sugar to the agar-agar solution and remove from heat. Then add the soaked gelatin (1) and allow to dissolve.
4. Add milk and vanilla extract to the agar-gelatin solution (3) (Figure 3.5C), mix well, pour into glasses (Figure 3.5D), and cool in the refrigerator or an ice-water bath.
5. Drain *nata de coco*.
6. Mix blueberry jam and Kirshwasser or *nata de coco* syrup in a bowl (Figure 3.5E).
7. Pour blueberry sauce (6) over the set milk jelly (4), and garnish with *nata de coco* (5) and mint leaves (Figure 3.5F).

* *Nata de coco* syrup is more suitable for children.

3.6.4 How to Make "*Kombucha*"

(Makes about 2 L)

1 *kombucha* culture (or scoby)*

30 mL (2 tbsp) cider vinegar*

2 L (8 cups) water

6–8 g (3–4 tbsp) tea leaves† or 3–4 tea bags

160 g white sugar

1. Boil 500 mL water in a pan and add tea leaves and continue to boil for 20–25 minutes.
2. Strain tea and stir in white sugar to dissolve.
3. Add 1.5 L water to tea and cool to room temperature before adding the *kombucha* starter (high heat will kill the starter).
4. Pour the cooled tea into a glass jar or bowl.
5. Place *kombucha* culture and cider vinegar carefully on the tea, cover with a towel, and secure the top of the glass jar with a rubber band.
6. Keep the glass jar in a warm, dark place (at 23°C–30°C) for at least 5 days to ferment.

pH = 2.80

3.6.5 Russian Salad Dressing (Figure 3.6)

(Makes about 450 g)

40 g (4½ tbsp) white sugar

45 mL (3 tbsp) water

180 g (⅘ cup) vegetable oil

140 g (a little under ½ cup) ketchup

30 mL (2 tbsp) lemon juice

---

* You can substitute 200 g *kombucha* from a previous batch for the *kombucha* culture and cider vinegar.

† You can use green or black tea.

FIGURE 3.6 RUSSIAN SALAD DRESSING. (A) DISPERSE SUGAR IN WATER. (B) HEAT THE DISPERSION TO THE SPUN-THREAD STAGE AND THEN COOL. (C) ADD BC AND OTHER FOOD ADDITIVES. (D) BLEND TO DISSOLUTION. (E) ADD VEGETABLE OIL WHILE STIRRING CONTINUOUSLY. (F) POUR SYRUP INTO THE MIXTURE. (G) WHISK WELL.

5 mL (1 tbsp) vinegar*
6 g BC powder
3 g (½ tbsp) salt
0.6 g (½ tbsp) paprika powder
4 g (1½ tbsp) celery seed
18 g (1 tbsp) Worcestershire sauce
3 g (½ tbsp) onion powder

1. Heat white sugar and water in a pan (Figure 3.6A) to the spun-thread stage (110°C), remove from heat, and then cool to room temperature (Figure 3.6B).

* You can use any vinegar you like.

2. Mix BC, salt, paprika, celery seed, and onion powder in a bowl. Then blend in ketchup, lemon juice, vinegar, and Worcestershire sauce (Figure 3.6C and D) (Hint 3.6.5.1); add vegetable oil gradually with stirring (Figure 3.6E) (Hint 3.6.5.2).
3. Pour the syrup (1) into the mixture (2) (Figure 3.6F) and whisk well (Figure 3.6G).

pH = 3.30

Preparation hints:

3.6.5.1 Strong mixing will prevent lumps from forming.

3.6.5.2 It is important for the salad dressing to emulsify. Adding oil gradually produces a good emulsion.

This dressing is good on fresh and boiled vegetables, boiled seafood, or fried fish. This recipe was actually invented in the United States.

3.7 TIPS FOR THE AMATEUR COOK AND PROFESSIONAL CHEF

- BC is a low-calorie additive.
- The use of BC in combination with other agents, such as sucrose and carboxymethylcellulose, improves dispersion of the product.
- BC is particularly recommended for use at low levels, to prevent flavor interactions and when stability over a wide pH range is required.

References and Further Reading

Brown, A. J. 1886. An acetic acid ferment which forms cellulose. *J. Chem. Soc.* 49: 432–9.

Castro, F. D., Sumague, J., and D. D. Villa. 1994. *How to Produce Nata de Coco.* Manila: Technology and Livelihood Resource Center. http://www.southpacificbiz.net/library/docs/howto/How%20to%20Produce%20Nata%20de%20Coco.pdf (searched on 1 Jan 2018).

Chao, Y. P., Sugano, Y., Kouda, T., Yoshinaga, F., and M. Shoda. 1997. Production of bacterial cellulose by *Acetobacter xylinum* with air-lift reactor. *Biotechnol. Tech.* 11:829–32.

Czaja, W., Romanovicz, D., and R. M. Brown Jr. 2004. Structural investigations of microbial cellulose produced in stationary and agitated culture. *Cellulose*. 11:403–11.

Jagannath, A., Raju, P. S., and A. S. Bawa. 2010. Comparative evaluation of bacterial cellulose (nata) as a cryoprotectant and carrier support during the freeze drying process of probiotic lactic acid bacteria. *LWT—Food Sci. Technol.* 43:1197–203.

Jonas, R., and L. F. Farah. 1998. Production and application of microbial cellulose. *Polym. Degrad. Stab.* 59:101–6.

Karahan, A. G., Kart, A., Akoglu, A., and M. L. Cakmakc. 2011. Physicochemical properties of low-fat soft cheese *Turkish Beyaz* made with bacterial cellulose as fat mimetic. *Int. J. Dairy Technol.* 64:502–8.

Lin, K.-W., and H.-Y. Lin. 2004. Quality characteristics of Chinese-style meatball containing bacterial cellulose (*Nata*). *J. Food Sci.* 3:107–11.

Nussinovitch, A. 1997. *Hydrocolloid applications. Gum technology in the food and other industries.* London: Blackie Academic & Professional.

Nussinovitch, A. 2003. *Water soluble polymer applications in foods.* Oxford: Blackwell Publishing.

Nussinovitch, A., and M. Hirashima. 2014. *Cooking innovations, using hydrocolloids for thickening, gelling and emulsification.* Boca Raton, FL: CRC Press, Taylor & Francis Group.

Omoto, T., Uno, Y., and I. Asai. 2000. Bacterial cellulose. In *Handbook of hydrocolloids*, ed. G. O. Phillips and P. A. Williams, 724–40. Cambridge: Woodhead Publishing Limited.

Seeds of health, Fermenting, *Kombucha*. http://www.seedsofhealth.co.uk/fermenting/index_kombucha.shtml (searched on 1 Jan 2018).

Vandamme, E. J., De Baets, S., Vanbaelen, A., Joris, K., and P. De Wulf. 1998. Improved production of bacterial cellulose and its application potential. *Polym. Degrad. Stab.* 59:93–9.

CHAPTER 4

Cereal β-Glucans

4.1 INTRODUCTION

Oats and barley contain high concentrations of mixed-linkage β-D glucans, which are hydrocolloid polysaccharides with either β-1,3 or β-1,4 linkages. The human digestive system cannot digest the mixed linkages. As a result, the inclusion of oats, barley, or highly concentrated β-glucans in food products has garnered interest as a means of enhancing fiber consumption. The United States Food and Drug Administration (FDA) has recognized the significance of soluble oat fiber (designated as β-glucan) from oat flour and bran by permitting beneficial health claims on food labels. To gain health benefits from cereal β-glucans, a highly concentrated formulation should be prepared for incorporation into the desired food products, so as to avoid changes in that product's sensory characteristics. The structure of β-glucan influences its functionality, and genetic modification of cereal β-glucan's structure may expand its potential human health benefits. Nevertheless, as the acceptability of genetically modified products remains low and uncertain, plant breeding will surely play a role in the production of cereal β-glucans having novel structures, potentially elevated yields, and more far-reaching functionality.

4.2 STRUCTURE

The primary structure of mixed-linkage cereal β-glucans is a linear chain of glucopyranosyl monomers linked by a mixture of single β-1,3 linkages and consecutive β-1,4 linkages. The composition of β-glucans from cereal grains can be determined by reaction with lichenase and characterization by methylation analysis. Hydrolysis by lichenase helps determine the percentage of cellotriosyl and cellotetraosyl units in the β-glucan molecule. High-performance liquid chromatography (HPLC) can be used to determine the molar ratio of tri- to tetrasaccharides in cereal β-glucans. High-performance size-exclusion chromatography (HPSEC) can be helpful in determining the molar mass of β-glucans. Molecular weights of β-glucan vary widely, with those from oat being $0.065 - 3 \times 10^6$, barley $0.15 - 2.5 \times 10^6$, and wheat $0.25 - 0.7 \times 10^6$. The content of β-glucan in cereal grains can be measured by diagnostic kits. Better accuracy can be achieved using murine monoclonal antibodies and enzyme-linked immunosorbent assay (ELISA), which can detect nanogram quantities of soluble β-glucan extracted from cereal flour.

4.3 EXTRACTION AND PURIFICATION

Many factors influence the solubility, extractability, and yield of oat β-glucans. These include particle size, pretreatment of cereal grains, and extraction conditions (i.e., temperature, pH, solvents, and ionic strength). Dry milling of cereals enriches the β-glucan content. Impact- or roller-milling followed by sieving of oat groats effectively retains bigger particles, increasing β-glucan content per unit weight by 1.7-fold. Tempering oat groats under 12% moisture for 20 minutes, prior to roller-milling, enhances bran yield by 2-fold. The allocation of β-glucan inside cereal grains differs with variety and, as a result, the milled portions might differ in their β-glucan contents. Various parameters, such as growing conditions, location, and season, have considerable effects on β-glucan content of hull-less barley cultivars, but they have no effect on β-glucan allocation inside the kernel. Numerous enzymatic methods have been developed to produce useful oat β-glucan-enriched products.

In addition, several enzymatic procedures have been combined with solvent extraction to enrich for fiber components. Enrichment can also be achieved when alkaline or acidic conditions are combined with elevated temperatures, dry-milling, enzymes, and solvents.

β-Glucan solubility can be enhanced by increasing the water temperature from 20°C to 80°C. Enzyme-deactivated and untreated oat bran concentrate can be extracted with sodium carbonate solution at pH 10.0 and 40°C. Oat β-glucan is then isolated by alcohol precipitation, dialysis, or ultrafiltration. These methods can produce an oat product with 60%–65% β-glucan, but the resultant molecular weights and viscosities vary with processing conditions.

4.4 HEALTH CLAIMS

Many studies have shown that cholesterol lowering is an important health benefit of β-glucan consumption. However, some studies have shown that these effects decrease upon incorporation of hydrocolloids into foods. This may be due to the unfavorable effects of food processing on oat β-glucan's hypocholesterolemic function or due to the food matrix, or a combination of the two. Moreover, in some cases, the effect of barley β-glucan on lipid profile has been found to be variable among subjects. Yet other studies have shown that cereal β-glucans can offer physiological health advantages to a broad range of demographics, irrespective of gender, society, or age. Cereal β-glucans decrease foods' glycemic index. Glycemic indices of patients eating a prototype β-glucan-enriched cereal item for consumption were markedly lower than in those eating a commercial oat-bran cereal. Metabolic control of diabetes by cereal β-glucans is almost certainly due to their inherently high viscosity, which delays nutrient absorption. Oat β-glucans are able to delay gastric emptying, reduce nutrient absorption, affect motility in the small bowel, and prolong postprandial satiety. Chemical modifications of cereal β-glucans have been reported to accelerate their health-beneficial physiological effects. In addition, oat β-glucan can improve blood pressure. Finally, it has been suggested that oat-based products might have negative effects on mineral absorption, possibly as a result of high phytate content in grains.

4.5 MARKETABLE PRODUCTS

Oatrim is one of the earliest commercial oat β-glucan products. It was developed and patented by the United States Department of Agriculture (USDA) in the early 1990s. It is currently marketed as Beta Trim® and distributed by Skidmore Sales and Distributing, Inc., West Chester, OH. Oatrim is an effective fat replacer, it lowers glycemic index in humans, and its powder or gel can be used in meats, dairy products, frozen desserts, salad dressings, mayonnaise, sauces, soups, margarine, baked goods, and beverages. The USDA improved on Oatrim by producing Nutrim-OB, which has ~12%–13% β-glucan, and C-trim, which has 20%–50% β-glucan content. Nutrim-OB is manufactured and marketed by Van Drunen Farms, Momence, IL. Nutrim-OB can serve as a replacement for fat, conferring moistness, softness, and cohesiveness to baked goods, cheeses, and Asian foods.

Three oat β-glucan hydrocolloids, named C-trim20, C-trim30 and C-trim50 (corresponding to 20%, 30%, and 50% β-glucan contents, respectively), were also prepared by the USDA. These products differ from Nutrim-OB by undergoing a supplementary centrifugation step prior to jet-cooking to decrease the starch content and design the composition of the final product. C-trim oat products are currently produced and marketed by Van Drunen Farms.

OatWell® 14%, 16%, and 22% was developed by the Swedish Oat Fibre AB, Varobacka, Sweden. OatWell® is a light yellow to brown, neutral-flavored product. It can be included in cereal bars, snacks, breakfast cereals, breads, and other baked goods. Nuture® 1500 with 15% β-glucan is manufactured by GTC Nutrition (Missoula, MT), and Natureal®, which has 20% β-glucan, is manufactured by Finn Cereal, Vantaa, Finland. The former can be incorporated into fruit juices, skimmed milk, and other beverages, as well as in baked goods. The latter is well suited for beverage powders, baked goods, and confections, among many other products. Viscofiber® is an enzymatically produced concentrated oat or barley β-glucan product. It was developed by Cevena Bioproducts, Inc., Edmonton, Canada. This high-viscosity product has a light color and a bland flavor and is suitable for food applications such as nutrition bars, beverages, soups, breakfast cereals, baked goods, and dairy products.

Glucagel® is produced by Gracelinc, Ltd., Christchurch, New Zealand. It contains 75%–89% pure β-glucan and dissolves readily in water at up to 20% at temperatures >80°C. This product is achieved through an enrichment process with a freeze-thaw step. The product forms soft, thermoreversible gels that set and melt at ~60°C. Cerogen™ was developed by Roxdale Foods Ltd., Auckland, New Zealand. It can make use of either oats or barley for β-glucan enrichment. The product contains 70%–90% β-glucan, and is water-soluble at room temperature, producing a soft, translucent gel. It has been reported to lower the glycemic response in healthy adult males.

4.6 FOOD APPLICATIONS AND REGULATORY STATUS

Cereal β-glucans can be included in a variety of food products, such as baked goods (muffins and cakes), pasta, noodles, muesli cereals, milk products, soups, salad dressings, beverages, and reduced-fat dairy and meat products. The qualities that they confer, such as breadmaking performance, water-binding and emulsion-stabilizing capacity, thickening ability, texture, and appearance, seem to be related to the concentration, molecular weight, and structure of the polysaccharide. Glucagel® incorporated into white wheat bread at 2.5%–5% concentration significantly decreased the height and loaf volume, as well as the rate of reducing sugar release. Cookies had reduced spread and increased elasticity and water content after the inclusion of 10% C-trim20 or C-trim30.

Incorporating barley β-glucans into durum wheat pasta resulted in a greater solid loss during cooking, decreased hardness, and increased swelling. Barley β-glucan-enriched pasta was darker than durum wheat pasta, but had exceptional cooking characteristics, with regards to adhesiveness, bulkiness, and firmness. Oat β-glucan incorporated into Asian noodles successfully reduced the use of rice flour by 50%, while maintaining noodle firmness and taste. To obtain oat bran-loaded maize spaghetti with sensory properties similar to those of the bran-less spaghetti, two consecutive trials were performed. In the first trial, the amount of oat bran added to the spaghetti was sequentially increased until the overall sensory quality of the pasta reached the set sensory threshold (at an oat bran concentration of 20%). The second trial was aimed at improving the overall sensory quality of oat bran-loaded maize spaghetti. In particular, an attempt was made to increase the sensory quality of spaghetti with 20% oat bran inclusion by means of structuring agents. The effects of different kinds of hydrocolloids and egg white powder on the rheological properties of the dough, as well as on the quality attributes of the pasta, were examined. Rheological analysis demonstrated that the addition of hydrocolloids and egg white to dough enriched with 20% oat bran does not cause any substantial difference in viscoelastic properties, compared to samples without any structuring agents. The best overall quality for both fresh and dry spaghetti was obtained by the addition of carboxymethylcellulose and chitosan (Chapter 5) at a concentration of 2%.

Oat β-glucan was favored over barley β-glucan in some dairy products. Improved probiotic viability and enhanced propionic and lactic acid production in yoghurt were described. A new product was introduced in the Swedish market, consisting of a 200-mL cup of low-fat yoghurt, with a separate cup of muesli attached to it. The muesli contained 4 g of oat β-glucans per serving. In clinical tests, the test meal significantly lowered blood sugar and insulin responses, compared to a reference meal without oat β-glucans. Oat β-glucans have been described as fat replacers in cheeses. In these cases, producers are trying to retain the textural and sensory qualities of the original product. The physical properties of 3-month-old, low-fat Cheddar cheeses produced with a β-glucan hydrocolloid component, named Nutrim (nutraceutic substitution of fat), were studied. In general, it was concluded that significant substitution of fat with the Nutrim component results in softer Cheddar cheeses characterized by decreased melting time and sensory properties. The challenge of utilizing cereal β-glucans as fat replacers in cheese lies in producing a final product with unaltered sensory attributes.

Soups that included oat bran were reported to be more viscous, with a more intense flavor. Barley β-glucan can produce a gelled network that stabilizes whey protein concentrate foams and emulsions. Inclusion of 0.3% β-glucan in sausages reduced the fat content. Higher inclusions had less favorable effects on the texture of sausage. Inclusion of β-glucans in different beverages retained the original texture of the product, but higher the inclusion, less desirable the reported results. Daily intake of a beverage containing 5 g of oat β-glucan significantly lowered the total serum cholesterol concentrations by 7.4%, compared to a control beverage containing rice starch. Oat products have shown significant medicinal benefits in humans. Mixed-linkage β-1,3–β-1,4 D-glucans have been suggested as the active constituents in oats. Because of the high quality demanded by the food industry, alternate uses have to be developed for those oat varieties, which do not meet the requirements of the regulatory bodies and are, thus, considered unfit for consumption. One such option is to use oats as a feedstuff for pigs. However, β-glucans are known to interfere with digestion and absorption, mainly in young pigs. Four oat-based diets, containing different levels of β-glucans (4.1%, 3.3%, 2.1%, and 1.6%), were fed to finisher pigs (52–107 kg body weight). Their growth performance and carcass and pork quality were compared to those of pigs fed with a standard wheat–barley-based diet. No evidence of serious dilatory effects of the β-glucans was detected, providing support to the decision to include oats in finisher pig diets. A few studies mention the option of creating food films from different β-glucans. Barley films that remain intact after 24 hour immersion in water might find interesting uses in the near future. All cereal β-glucan-enriched products are generally regarded as safe (GRAS), because they are natural constituents of cereal grains. A few of them have gained recognized kosher status as well as approval for definition as an "organic natural fiber".

4.7 RECIPES

4.7.1 *LASSI* (FIGURE 4.1)

(Serves 5)

5 g (2½ tbsp) cereal β-glucan powder

500 g (2 cups) yoghurt

200 mL (⅘ cup) milk

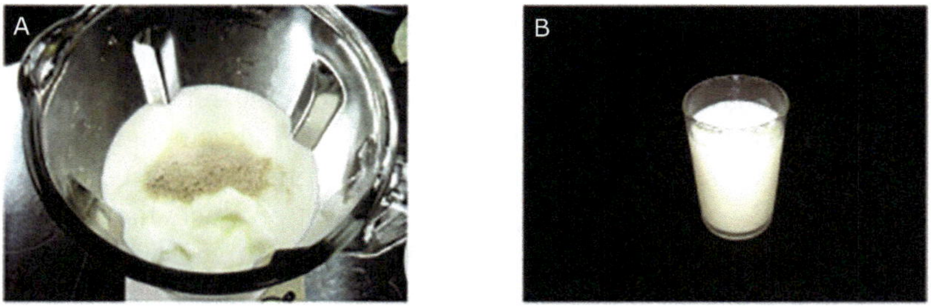

FIGURE 4.1 *LASSI*. (A) BLEND INGREDIENTS. (B) SERVE.

200 g ice
80 g (a little under 4 tbsp) cane sugar syrup

1. Put all ingredients in a blender (Figure 4.1A), and blend in two 5-second pulses (Hint 4.7.1.1).
2. Pour into glasses (Figure 4.1B).

pH = 4.20

Preparation hints:
4.7.1.1 Blending for a longer time results in separation of fat and water in the yoghurt.

This recipe originates from India. You can add sugar and any other type of fruit to it; mango *lassi* is very popular in India. Cereal β-glucan is added for dietary fiber intake.

4.7.2 BASQUE OMELET (FIGURE 4.2)

(Serves 5)
500 g (10) eggs
200 g (5) sweet pimento (green and red)
200 g (1) onion
8 g (a little over ½ tbsp) 0.4% cereal β-glucan solution
Salt, *piment d'Espelette* or paprika, olive oil

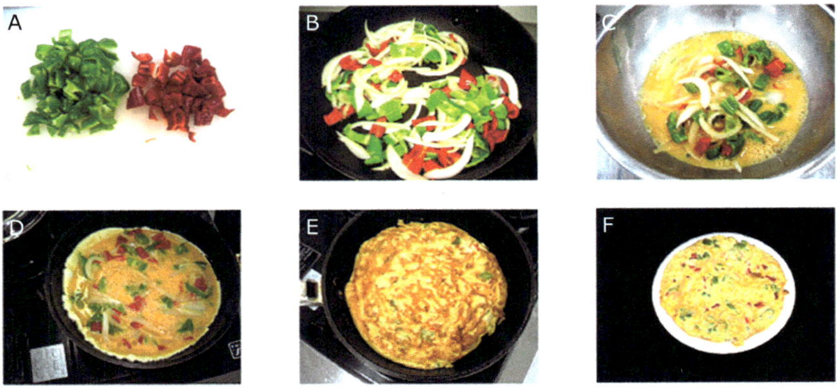

FIGURE 4.2 BASQUE OMELET. (A) CUT PIMENTO INTO BITE-SIZED PIECES. (B) STIR-FRY CUT PIMENTO AND SLICED ONION UNTIL SOFT. (C) BEAT EGGS, ADD SALT, *PIMENT D'ESPELETTE*, CEREAL β-GLUCAN SOLUTION, AND THE STIR-FRIED PIMENTO AND ONION. (D) POUR INTO A HOT PAN WITH OLIVE OIL. (E) TURN OVER WHEN BOTTOM IS READY. (F) SERVE.

1. Cut pimento in two lengthwise, remove seeds and core (Hint 4.7.2.1), and cut into bite-sized pieces (Figure 4.2A).
2. Slice onion thinly.
3. Stir-fry cut pimento (1) and sliced onion (2) until soft (Figure 4.2B) and remove from heat.
4. Beat eggs in a bowl, and add salt, *piment d'Espelette*, cereal β-glucan solution, and the stir-fried pimento and onion (3) (Figure 4.2C).
5. Heat olive oil in a frying pan (Hint 4.7.2.2), and pour in egg dispersion (4) (Figure 4.2D).
6. When the bottom of the omelet (5) is done, turn it over using a plate (Hint 4.7.2.3), and heat the second side for 2 minutes (Figure 4.2E and F).

pH = 7.32

pH of 0.4% cereal β-glucan solution = 7.64

Preparation hints:

4.7.2.1 Seeds and core are very hot and spicy.

4.7.2.2 The more oil you use, the easier it is to flip the omelet over during cooking.

4.7.2.3 Slide the half-done omelet into a plate; turn frying pan upside down and place it on the plate with the omelet. Holding the plate and frying pan together, turn over and remove plate.

This is a traditional Basque dish. It is good hot or cold. Cereal β-glucan is added as a texture modifier, to make the omelet fluffy.

4.7.3 Boiled Mackerel with Vinegar (Figure 4.3)

(Serves 5)

500 g (5 pieces) mackerel
100 g (1) long onion or leak
50 g ginger
120 mL (a little under ½ cup) vinegar
130 mL (a little under ½ cup) soy sauce
60 g (a little under ½ cup) white sugar
15 g (1 tbsp) 0.4% cereal β-glucan solution
120 mL (a little under ½ cup) Japanese broth or water

Figure 4.3 Boiled mackerel with vinegar. (A) Wash fish and pat dry. (B) Slice ginger and long onion. (C) Combine ginger, onion, vinegar, soy sauce, white sugar, cereal β-glucan solution, and Japanese broth and bring to a boil. (D) Add fish to the pan and simmer over medium heat. (E) Serve.

1. Wash mackerel, and pat dry with a paper towel (Figure 4.3A).
2. Slice ginger, and cut long onion in 5-cm pieces (Figure 4.3B).
3. Put the sliced ginger and cut onion (2), vinegar, soy sauce, white sugar, cereal β-glucan solution, and Japanese broth in a pan, and bring to a boil (Figure 4.3C).
4. Add the mackerel (1) to the pan (3) (Figure 4.3D) (Hint 4.7.3.1), and simmer for 10 minutes over medium heat.
5. Plate the mackerel (Figure 4.3E).

pH of seasoned broth = 3.98

Preparation hints:

4.7.3.1 Fish can easily break into pieces or lose its shape during boiling. Do not turn fish over while boiling.

This recipe is full of refreshing tastes. Cereal β-glucan removes the fishy smell of mackerel.

4.8 TIPS FOR THE AMATEUR COOK AND PROFESSIONAL CHEF

- Spreading a β-glucan dispersion over fish or meat before baking or stir-frying gives it a tender, fluffy texture as well as a glossy surface.
- A fluffy texture can be achieved by adding a β-glucan dispersion to omelet or pancake batter.
- Cooking time is reduced by 10%–30% using a β-glucan dispersion.
- Vegetables will stay fresh longer when β-glucan dispersion is added.

REFERENCES AND FURTHER READING

Biorklund, M., van Rees, A., Mensink, R. P., and G. Onning. 2005. Changes in serum lipids and postprandial glucose and insulin concentrations after consumption of beverages with β-glucans from oats or barley: a randomised dose-controlled trial. *Eur. J. Clin. Nutr.* 59:1272–81.

Duss, R., and L. Nyberg. 2004. Oat soluble fibers (β-glucans) as a source for healthy snack and breakfast foods. *Cereal Foods World* 49:320–5.

Fortin, A., Robertson, W. M., Kibite, S., and S. J. Landry. 2003. Growth performance, carcass and pork quality of finisher pigs fed oat-based diets containing different levels of β-glucans. *J. Anim. Sci.* 81:449–56.

Havrlentova, M., Petrulakova, Z., Burgarova, A., Gago, F., Hlinkova, A., and E. Šturdik. 2011. Cereal β-glucans and their significance for the preparation of functional foods—a review. *Czech. J. Food Sci.* 29: 1–14.

Inglett, G. E. 2000. Soluble hydrocolloid food additives and method of making. U.S. Patent No. 6,060,519.

Inglett, G. E., Peterson, S. C., Carriere, C. J., and S. Maneepun. 2005. Rheological, textural, and sensory properties of Asian noodles containing an oat cereal hydrocolloid. *Food Chem.* 90:1–8.

Keenan, J. M., Pins, J. J., Frazel, C., Moran, A., and L. Turnquist. 2002. Oat ingestion reduces systolic and diastolic blood pressure in patients with mild or borderline hypertension: a pilot trial. *J. Family Practice* 51:369.

Knuckles, B. E., Chiu, M.-C. M., and A. A. Betschart. 1992. β-glucan-enriched fractions from laboratory-scale dry milling and sieving of barely and oats. *Cereal Chem.* 69:198–202.

Lazaridou, A., Biliaderis, C. G., and M. S. Izydorczyk. 2003. Molecular size effects on rheological properties of oat beta-glucans in solution and gels. *Food Hydrocolloids* 17:693–712.

Lee, S., Inglett, G. E., and C. J. Carriere. 2004. Effect of Nutrim oat bran and flaxseed on rheological properties of cakes. *Cereal Chem.* 81:637–42.

Luhaloo, M., Martensson, A.-C., Andersson, R., and P. Aman. 1998. Compositional analysis and viscosity measurements of commercial oat brans. *J. Sci. Food Agric.* 76:142–8.

Malkki, Y., and E. Virtanen. 2001. Gastrointestinal effects of oat bran and oat gum a review. *Lebensm. Wissn u-Technol.* 34:337–47.

Nussinovitch, A. 2003. *Water soluble polymer applications in foods.* Oxford: Blackwell Publishing.

Nussinovitch, A., and M. Hirashima. 2014. *Cooking innovations, using hydrocolloids for thickening, gelling and emulsification.* Boca Raton, FL: CRC Press, Taylor & Francis Group.

Padalino, L., Mastromatteo, M., Sepielli, G., and M. A. Del Nobile. 2011. Formulation optimization of gluten-free functional spaghetti based on maize flour and oat bran enriched in β-glucans. *Materials* 4:2119–35.

Pomeroy, S., Tupper, R., Cehun-Aders, M., and P. Nestel. 2001. Oat beta glucan lowers total and LDL-chlesterol. *Aust. J. Nutr. Diet* 58:51–5.

Potter, R. C., Fisher, P. A., Hash, K. R., and J. D. Neidt. 2003. Method for concentrating beta-glucan film. U.S. Patent No. 6,624,300.

Redmond, M. J., and D. A. Fielder. 2005. Oat extracts: refining, compositions and methods of use. U.S. Patent Application 20,050,042,243.

Stevenson, D. G. 2009. Cereal β-glucans. In *Handbook of hydrocolloids*, 2nd edn., ed. G. O. Phillips and P. A. Williams, 615–39. Boca Raton: CRC Press and Oxford: Woodhead Publishing Ltd.

Tapola, N., Karvonen, H., Niskanen, L., Mikola, M., and E. Sarkkinen. 2005. Glycemic responses of oat bran products in type 2 diabetic patients. *Nutr. Metab. Cardiovasc. Dis.* 15:255–61.

Vasanthan, T., and F. Temelli. 2005. Grain fiber compositions and methods of use. U.S. Patent Application 200,050,208,145.

Wood, P. J., Siddiqui, I. R., and D. Paton. 1978. Extraction of high viscosity gums from oats. *Cereal Chem.* 55:1038–49.

CHAPTER 5

Chitin and Chitosan

5.1 INTRODUCTION

The term "chitin" comes from the Greek word *chiton*, meaning coat of mail, so named because of its function as a protective coating for invertebrates. Chitin occurs in substantial quantities in the cell walls of numerous lower plants. It is not soluble in water; nevertheless, a number of its derivatives or chemical modifications are. Their dissolution produces solutions with high viscosity which, in some cases, can gel. Major sources of chitin are shrimps, crabs, and lobster shells. They are gathered in large quantities when these comestibles are processed. Chitin is a high-molecular-weight linear polymer of 2-acetamido-2-deoxy-D-glucopyranosyl units. Similar to cellulose, the monomer units are linked by β–D-(1→4) bonds. Chitin and chitosan are commercially available in a variety of grades. Chitin isolates differ from each other in many aspects, including degree of acetylation, elemental analysis, molecular size, and polydispersity. Chitin and chitosan are not present in human tissues, but acetylglucosamine and chitobiose are found in glycoproteins and glycosaminoglycans. Since chitosan is biodegradable, non-toxic, non-immunogenic, and biocompatible in animal tissues, much research into its use in medical applications has been performed.

5.2 SOURCE

Chitin is the most plentiful organic ingredient in the skeletal matter of invertebrates. It is found in arthropods, annelids, and mollusks, supplying skeletal support and body armor. The most important and economic source of chitin is crustaceans, which are gathered on a large scale as food. Today, freezing and canning of lobsters, crabs, and shrimps yield considerable quantities of crustacean waste materials. Chitin can be extracted from these wastes, which include mostly heads and shells. Hard crustacean shells are composed of ~15%–20% chitin and as much as 75% calcium carbonate, in addition to skeletal protein. The less hard shells from shrimps are comprised of 15%–30% chitin and 13%–40% calcium carbonate, in addition to skeletal protein. Another potential source of chitin is fungi, as it is a known component of many fungi's mycelia and spores. Chitin can also be found in the exoskeleton of most insects where it comprises up to 60% of the flexible skeletal portions. The average amount of chitin in the cuticle of various species has been assessed at 33%. The cuticle consists of alternating layers of protein and chitin impregnated with calcium carbonate and pigments and mixed with polyphenols.

5.3 PREPARATION AND AMOUNT AVAILABLE

Although shells can be extracted whole, shells ground to particle sizes of between 1 and 6 mm are preferred. The shell pieces are treated with 5% hydrochloric acid via countercurrent extraction. About 24 hours are needed for demineralization to reduce the ash content of the shell to 0.4%–0.5%. A further step includes deproteinization, accomplished by treating the mineralized shells with trypsin or pepsin.

Deproteinization by alkali can be used, especially when deacetylated chitin (chitosan) is required. Alkali deproteinization is performed by blending the demineralized ground shells with three successive quantities of 5% sodium hydroxide at 85°C–90°C. The duration of each treatment is 30–45 minutes, and each treatment is followed by a short water wash to eliminate the incompletely solubilized remains. At this stage, chitin is light pink in color due to the presence of residual pigments. These residual pigments can be removed by solubilization by mild oxidation in an acidified hydrogen peroxide solution for 6–7 hours at an ambient temperature.

Removal of N-acetyl groups from chitin for the production of chitosan requires the use of concentrated alkali. A characteristic deacetylation bath consists of, for example, 2 parts sodium hydroxide, 1 part 95% ethanol, and 1 part ethylene glycol. This solution is heated to 120°C, after which the alcohol refluxes. The deacetylation reaction causes some alkaline cleavage of the polysaccharide, resulting in reduced viscosity. The reaction is completed when the acetyl content is reduced to low levels. An approximate yield of chitosan from raw crab shells is 7%. The viscosities of good-quality, medium-grade, and low-grade preparations are 1,200, 160, and 15 mPa s, respectively, when deacetylated chitin is present at a concentration of 1.25%. Commercial deacetylated chitin is 80%–85% deacetylated. It also contains 6.5% nitrogen. About a hundred thousand tons of shrimps, lobsters, and crabs are processed in the United States. Large sources of shells are also processed worldwide.

5.4 STRUCTURE

Chitin is composed of N-acetyl-D-glucosaminyl units that are linked together just like the D-glucopyranosyl units in cellulose. Molecular weights of ~143–210 kDa suggest that the molecule is a highly linear array of monomer units. The regularity of the structure, the large number of hydrogen bonds, and the dipole interactions probably account for the insolubility of natural chitin. Chitin that has passed through deacetylation produces a cationic polysaccharide. Other soluble derivatives, such as hydroxyethyl, have become useful in many applications, such as wet-end additives in paper-making and as precipitants for anionic wastewater from meat-packing plants.

5.5 PROPERTIES OF CHITOSANS AND DERIVATIVES

A variety of chitin derivatives can be produced in a procedure similar to that for cellulose derivatives. The chitin derivatives are recognized for their remarkable attributes and prospective applications. The insolubility of chitin can be overcome with the support of hypochlorite as 6-oxychitins. β-Chitin is soluble in calcium chloride-saturated methanol, among other solvents. Under some circumstances, chitin and chitosan can present hydrophilic water-swellable hydrogels. Gel creation is also supported by cross-linking agents or organic solvents, mainly for chitosan derivatives. Chitosan is soluble in aqueous acidic media, whereas other polysaccharides are usually neutral or anionic. Chitosan is similarly soluble in alkaline solutions at pH 10.0, where it displays glycosylamine functions. The creation of trade-appropriate derivatives, having good solubility in a variety of organic solvents, can be achieved through the introduction of hydrophobic constituents, by the means of acylation with long-chain fatty acyl halides or anhydrides. In an aqueous solution, where the polymer concentration is higher than a certain threshold level, intermolecular hydrophobic interactions lead to the formation of associations. As an outcome, these copolymers display thickening properties comparable to those achieved with higher-molecular-weight homopolymers and might play a vital role in waterborne technologies. Chitosan gels can be used for many non-food purposes as well. Gel formation can be induced by the complexation of beads in pentasodium tripolyphosphates or polyphosphoric acid via ionotropic cross-linking or interpolymer complexation. Chitin gel can be formed by treating chitosan-acetate salt solution with carbodiimide to restore acetamido groups. Chitosan gels can also be prepared using large organic counterions. Many other gel preparations, in amino acid solutions of ~pH 9.0, via enzymatic reactions, or via reactions with aldehydes, can be located in the scientific literature. Furthermore, information regarding chitosan macromolecular complexes and their creation and potential uses can be found elsewhere.

5.6 APPLICATIONS

5.6.1 Food Applications

Chitosan exhibits antimicrobial activity against a range of foodborne microorganisms and as a result, was evaluated for its ability to serve as a potential natural food preservative. Its bitter taste limits its use in meat and meat products, although it has the ability to control rancidity. There are two types of antioxidants: primary and secondary. The first are characterized by phenolic groups in the molecules and react during the earlier steps of the oxidative reaction. The second can chelate the metal ions that are involved in catalyzing the oxidative reaction. There is a large body of research documenting the effectiveness of chitosan as a secondary antioxidant.

Muscle food products are highly susceptible to the development of off-flavors and rancidity caused by oxidation of their highly unsaturated lipids. A warmed-over flavor develops in cooked poultry and uncured meat during storage, and there is a deterioration of the desired meaty flavor. A study of the effect of chitosan treatment on oxidative stability of beef concluded that its addition at 1% results in a 70% decrease in the 2-thiobarbituric acid values of meat after 3 days of storage at 4°C. A small quantity of added chitosan (0.05%, 0.1%, and 0.5% w/w) incorporated into a Turkish sausage mix had a positive effect on both the microbiological and sensory qualities of the product. Nevertheless, addition of greater amounts (such as 1%) negatively affected sensory quality and created problems in the processing itself. Incorporation of 0.1% chitosan of varying molecular weights into reduced-fat (~22%) Chinese-style sausage resulted in similar or better quality products, with respect to physicochemical, microbial, and sensory characteristics, with no adverse effects on the textural properties. Appropriate supplementation levels need to be established and the cost of chitosan justified before practical utilization in meat products. Effects of lower-molecular-weight chitosan (<150 kDa), degree of deacetylation (<85%), or pH of the chitosan solution on the properties, particularly antibacterial activity, of meat products, however, have not been thoroughly elucidated and require further research. The influence of chitosan on color stability and lipid oxidation in refrigerated ground-beef patties is packaging-specific. Use of 1% chitosan improves the surface red color of beef patties in carbon monoxide and aerobic packaging, but not in high-oxygen and vacuum packaging. While chitosan decreases lipid oxidation, it also minimizes the variations in lipid oxidation due to packaging. Packaging chitosan-treated patties in carbon monoxide modified atmosphere packaging improves color stability. The beef industry could exploit the synergistic effect of natural antioxidants and modified atmosphere packaging systems to minimize discoloration-induced sales loss.

When chitosan was mixed with essential oils, such as mint oil, the mixture showed an efficient scavenging of superoxide and hydroxyl radicals and effectiveness against Gram-positive bacteria. Chitosan films plus either mint or oregano essential oils were beneficial for shelf-life extension of pork cocktail salami or bologna slices. In other words, combinations of edible chitosan films plus essential oils could be effective against different bacteria, such as *Listeria monocytogenes* and *Escherichia coli*. Chitosan films are tough, flexible, and tear-resistant; they also have favorable characteristics, such as permeability to gases and water vapor. Chitosan is suitable as a texturizing agent for perishable foods. An example for the use of high-viscosity solutions is in the preparation of oriental foods and *tofu*. The gel properties of *tofu* depend on the curdling agent and method employed for its preparation. For the same kind of tofu, the inclusion of chitosan augments the gel strength and storage stability. Chitosan slightly changes the water-holding capacity of the *tofu* gel structure. Chitosan's effect on the gel properties of *tofu* showed a strong relationship with its degree of deacetylation.

Processing of clarified fruit juices frequently involves the use of clarifying agents. Chitosan salts carry a strong positive charge and are valuable as dehazing agents. Chitosan is a good clarifying agent for grapefruit juice, either with or without pectinase treatment, and an exceedingly valuable fining agent for apple juice, producing zero-turbidity products with 0.8 kg/m^3 of chitosan. When chitosan glutamate is added to apple juice, it can protect against fungal spoilage. Chitosan has good affinity for polyphenolic compounds and thus might influence the color of white wines upon its addition. Chitosan has proven to be a useful emulsifier. Its emulsions are highly stable under changes in temperature and ageing. They are also stable to flocculation and coalescence. Chitosan and its derivatives can act as primary emulsifiers because they are amphiphilic polyelectrolytes and as secondary emulsifiers because they can increase the viscosity of the continuous phase. Chitin nanocrystals assist in the production of stable coalescing emulsions for a period of 1 month, even when the droplets are of relatively larger size. An increase in nanocrystal concentration leads to the formation of smaller droplets and higher stability. It was suggested that nanocrystals adsorb at the oil-water interface, building up an inter-droplet network and a chitin nanocrystal network in the continuous phase.

A variety of other uses are possible, for example, the presence of chitin throughout osmotic dehydration of banana slices at the smallest concentration of 0.01% (w/v) created a statistically weaker tendency toward lightness (whitish hue) reduction during dehydration, and its addition is strongly recommended to prevent adverse decreases in lightness during this process. The presence of chitin during osmotic dehydration did not alter the tendency toward decreasing polyphenol oxidase activity due to different dehydration conditions. Other uses include deacetylated chitin as a sizing agent for natural and synthetic fibers. It has also been recommended as a beater stock additive for paper and spinning bath additive for cellophane. It also demonstrates easy adhesion to glass and plastics. Chitosan can be used for the purification of waste water due to its high sorption ability. Both chitin and chitosan are capable of forming complexes with metal ions, and this property has been utilized in Japan for water purification. The NH_2 group of chitosan is of interest due to its ability to form coordinate covalent bonds with metal ions. Chitosan powder and dried films have more potential use in metal ion complexing because they discharge most of their free amino groups at a pKa that is higher than that of the NH_2 groups of chitosan.

5.6.2 Nutritional and Health Aspects

More than 2,000 edible species, such as crustaceans and insects that are utilized as food, serve as sources of chitin. Chitinolytic enzymes are required for chitin digestion. Chitinases are present in plant foods, and consequently, the vegetables and fruits consumed by humans might assist in the digestion of chitin. Nevertheless, chitinases have been found in quite a few human tissues and their function has been associated with defense against parasitic infections and with some allergic conditions. In the excreta of physically fit individuals, there is a bacterial species that is able to hydrolyze chitin to N-acetylglucosamine; *Clostridium paraputrificum* in the human colon can synthesize and secrete chitinases that can take part in the digestion of chitin. An inherited deficiency in chitotriosidase activity is frequently reported in the plasma of Caucasian subjects, whereas this deficiency is rare in African populations. This may be an indication of the presence of activated macrophages in the milk of African women, and it is related to their chitin-consumption habits.

Chitosan is generally regarded as safe (GRAS), and therefore, the United States, Japan, and South Korea endorse the sale of chitosan-based nutraceuticals over the counter to control obesity, hypercholesterolemia, and hypertension. Chitosans have been utilized to prepare dietary supplements, but there is no standard for any of their applications; in other words, although several grades are achievable, there is no regulation or recommendation for adopting any particular grade for a particular use. Chitosan has been shown to lower serum cholesterol in humans, as evidenced by increased fecal steroid excretion, with no link to fat excretion. In addition, uptake of bile salts by chitosan-alginate gel beads has been reported. Chitosan cinnamate and analogous compounds were able to adsorb bile acids and release cinnamate. These results suggest that such compounds might be of interest for the field of complementary medicine because they can prevent lifestyle-related diseases. Chitosan preparations result in significantly greater weight loss, decreased total cholesterol levels, and decreased systolic and diastolic blood pressure when compared with placebo. Combinations of glucomannan, chitosan, fenugreek, and vitamin C were effective for losing body weight. Chitosan increases fecal fat excretion if consumed in sufficient quantities and might accelerate weight loss when combined with a low-calorie diet.

5.7 RECIPES

5.7.1 *TEMPEH* (FIGURE 5.1)

(Makes 1 sheet)
100 g soybeans
100 mL (2/5 cup) vinegar
1 g *ragi tempeh*
10 g (a little over 1 tbsp) potato starch
Water, a plastic ziplock bag

1. Soak soybeans in about 300 mL (1⅕ cups) of water and 50 g vinegar for 5–6 hours.
2. Strain the beans (1), and peel away the thin skin by hand.
3. Put 600 mL (2⅖ cups) of water, 50 g vinegar, and the soybeans (2) in a pot, and heat until soybeans are soft.
4. Strain soybeans (3), and cool to 40°C.
5. Mix *ragi tempeh* and potato starch in a bowl.
6. Put cooled soybeans (4) in another bowl, sprinkle the *ragi tempeh* mixture (5) over them, and mix in by hand.
7. Place the soybeans with *ragi tempeh* (6) in a ziplock bag, close bag, and poke holes in it with a toothpick.
8. Let bag sit at 30°C for 20 hours.

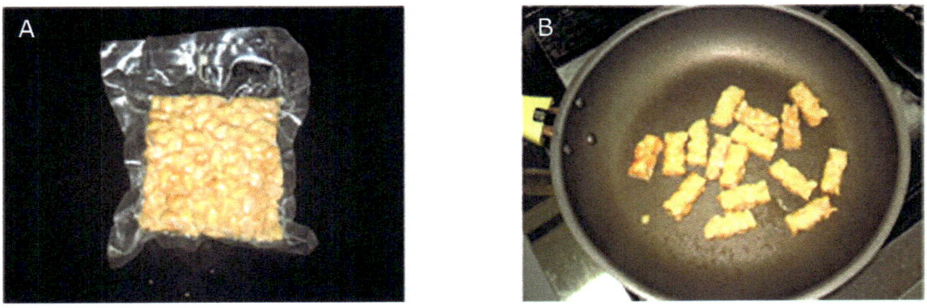

FIGURE 5.1 *TEMPEH*. (A) FERMENTED *TEMPEH* PRODUCT. (B) SAUTÉED *TEMPEH*.

Tempeh is one of Indonesia's soy products (Figure 5.1A). It is a fermented food that contains chitosan. *Tempeh* is usually eaten sautéed (Figure 5.1B). Deep-fried *tempeh* is also good, and it is quite tasty when used in salads.

There are a few examples using chitin and chitosan as thickeners. In the recipes that follow, chitosan is used as a preservative and dietary fiber.

5.7.2 Croquettes (Figure 5.2)

(Serves 5)
500 g potato
150 g minced pork

Figure 5.2 Croquette. (A) Mash boiled potatoes. (B) Stir-fry onion with salt and pepper and add pork. (C) Mix stir-fried onion and pork with mashed potato. (D) Form the potato mixture and allow to cool. (E) Mix wheat flour and chitosan. (F) Coat the formed potato with flour and chitosan. (G) Dip in beaten egg. (H) Coat with bread crumbs. (I) Deep-fry croquette until it becomes brown.

100 g onion
4 g (1 tsp) vegetable oil
Salt, pepper, wheat flour, chitosan, egg, bread crumbs, vegetable oil for deep-frying

1. Peel potatoes, and cut up roughly.
2. Boil cut potatoes (1), and mash (Figure 5.2A).
3. Chop onion finely, heat vegetable oil in a frying pan, and stir-fry the chopped onion, salt, and pepper until the onion becomes slightly transparent.
4. Add minced pork to the stir-fried onion (3) (Figure 5.2B), and continue to stir-fry until the minced pork is cooked; remove from heat.
5. Mix stir-fried onion and pork (4) into mashed potato (2) (Figure 5.2C).
6. Form the potato mixture (5) (Figure 5.2D), and cool (Hint 5.7.2.1).
7. Mix wheat flour and chitosan (Figure 5.2E); in a separate bowl, beat egg.
8. Coat the formed potato (6) with the flour and chitosan mixture (7) (Figure 5.2F), shake off excess flour, dip in beaten egg (7) (Figure 5.2G), and then coat with bread crumbs (Figure 5.2H).
9. Preheat vegetable oil to 170°C, and deep-fry the croquette (8) (Hint 5.7.2.2) until it becomes brown (Figure 5.2I).

Preparation hints:

5.7.2.1 Cool the potato mixture well so that the croquette does not fall apart during deep-frying.

5.7.2.2 Do not deep-fry too many croquettes at once because the temperature of the oil decreases sharply and the croquette may fall apart.

This recipe is for a standard croquette. You can add ingredients, such as corn or other vegetables, beef, chicken or fish, and spices such as black pepper, turmeric, and cumin.

5.7.3 Fried *Tofu* (Figure 5.3)

(Makes 500 g)

500 g (2 cups) soybean milk

80 g (⅓ cup) chitosan solution*

5 g (1 tbsp) bittern

Vegetable oil for deep-frying

1. Heat soybean milk to 40°C.
2. Heat chitosan solution to 40°C (Figure 5.3A).
3. Whisk together lukewarm chitosan solution (2) and lukewarm soybean milk (1) (Figure 5.3B) for 1–2 minutes, and heat to 80°C.
4. Add bittern to the soybean solution (3) and mix gently, and then let it stand for 10–15 minutes.
5. When the soybean curd (4) begins to coagulate, ladle it into molds covered with a piece of cloth, then press down the lid to remove water (see Chapter 18).
6. Remove *tofu* from the molds and put it in water.

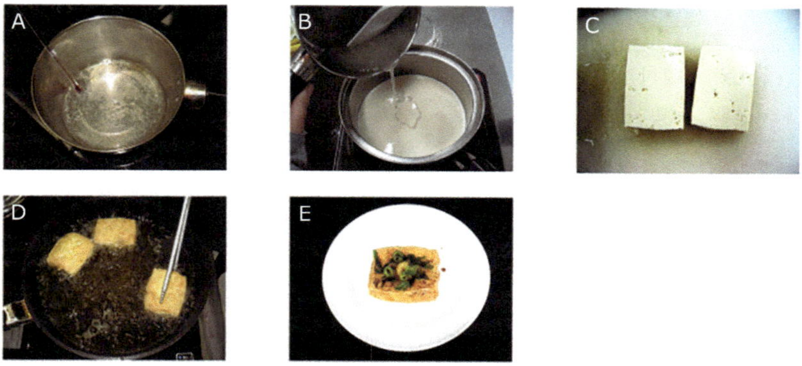

Figure 5.3 Fried *tofu*. (A) Heat chitosan solution. (B) Add chitosan solution to the soybean milk and stir while heating. (C) Slice coagulated and pressed *tofu*. (D) Deep-fry sliced *tofu* until it browns. (E) Serving suggestion: baked fried *tofu* with green onion and grated ginger.

* 5% chitosan powder, 10% vinegar, and 85% water. You can buy this readymade, and it is easy to use.

7. Cut *tofu* into 3–4 cm thick slices (Figure 5.3C).
8. Heat the vegetable oil to 180°C and deep-fry the cut *tofu* until it becomes brown (Figure 5.3D).

pH of fried *tofu* = 6.02

Chitosan helps the gelling agent (bittern) in this recipe. This recipe is eaten in various ways in Japan. Figure 5.3E shows baked fried *tofu* with green onion and grated ginger.

5.7.4 Carrot and Apple Juice (Figure 5.4)

(Serves 5)

5 g (a little over 1 tbsp) chitosan flakes
300 g (2) carrots
500 g (2) apples
30 mL (2 tbsp) lemon juice
50 g (2½ tbsp) honey
300 mL (1⅕ cups) water

1. Grind chitosan flakes in a coffee grinder (Figure 5.4A) (Hint 5.7.4.1).
2. Cut carrots and apples into small pieces.

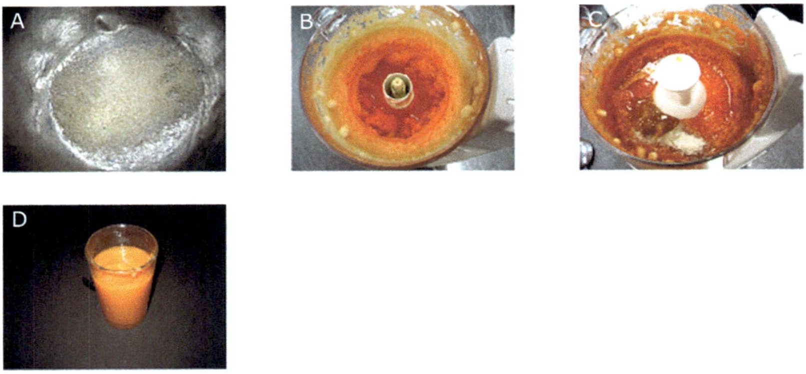

Figure 5.4 Carrot and apple juice. (A) Grind chitosan flakes. (B) Grind carrots and apples. (C) Mix ground chitosan, lemon juice, honey, and water with ground carrot and apple. (D) Serve.

3. Process the cut carrot and apple (2) in a food processor (Figure 5.4B).
4. Add the ground chitosan (1), lemon juice, honey, and water to the ground carrot and apple (3) (Figure 5.4C), and mix.
5. Pour the juice into glasses and serve (Figure 5.4D).

pH = 4.20

Preparation hints:

5.7.4.1 Grind chitosan flakes as finely as possible so that chitosan does not give an unpleasant texture.

This recipe contains much β-carotene and dietary fiber. It might have a beneficial effect on eye strain and intestinal disorders.

5.8 TIPS FOR THE AMATEUR COOK AND PROFESSIONAL CHEF

- Chitosan can be easily dissolved in an acetic acid solution or vinegar.
- Pickled vegetables with inclusion of 0.0125%–0.05% chitosan have a long shelf life with low concentrations of salt.
- Adding chitosan to deep-fried foods results in a crispier texture.
- The loaf volume of wheat bread can be increased by including up to 2% microcrystalline chitin in the formulation.
- Native chitosan has the potential to be used as a preservative in low-pH foods, either alone or in combination with other preservation systems.
- Addition of 2% chitosan increases both the gel strength and shelf life of *tofu*.
- Chitosan at a low concentration is effective in the clarification of different fruit juices.
- Incorporation of 1% chitosan minimizes premature browning in ground-beef patties stored in aerobic packaging, vacuum packaging, or carbon monoxide-modified atmosphere packaging.

References and Further Reading

Agullo, E., Rodrıguez, M. S., Ramos, V., and L. Albertengo. 2003. Present and future role of chitin and chitosan in food. *Macromol. Biosci.* 3:521–30.

Chang, K. L. B., Lin, Y.-S., and R. H. Chen. 2003. The effect of chitosan on the gel properties of tofu (soybean curd). *J. Food Eng.* 57:315–9.

Chatterjee, S., Chatterjee, S., Chatterjee, B. P., and A. K. Guha. 2004. Clarification of fruit juice with chitosan. *Process Biochem.* 39:2229–32.

Debritto, D., and O. B. G. Assis. 2007. A novel method for obtaining a quaternary salt of chitosan. *Carbohydr. Polym.* 69:305–10.

El Ghaouth, A., Arul, J., Asselin, A., and N. Benhamon. 1992. Antifungal activity of chitosan on post harvest pathogens: induction of morphological and cytological variations on *Rhizopurstolonifer*. *Mycol. Res.* 96:769–79.

Fukada, Y., Kimura, K., andY. Ayaki. 1991. Effect of chitosan feeding on intestinal bile acid metabolism in rats. *Lipids* 26:395–9.

Gades, M. D., and J. S. Stern. 2005. Chitosan supplementation and fat absorption in men and women. *J. Am. Dietetic Assoc.* 105:72–7.

Gokmen, M., and U. Gurbuz. 2011. Use of chitosan in turkish sausage (*Sucuk*) production and effects on quality. *Kafkas Universitesi Veteriner Fakultesi Dergisi* 17: S67–S71.

Gray, J. I., Gomaa, E. A., and D. J. Buckley. 1996. Oxidative quality and shelf life of meats. *Meat Sci.* 43:S111–23.

Ikeda, I., Sugano, M., Yoshida, K., Sasaki, E., Iwamoto, Y., and K. Hatano. 1993. Effects of chitosan hydrolysates on lipid absorption and on serum and liver lipid concentrations in rats. *J. Agric. Food Chem.* 41:431–5.

Kanatt, S. R., Chander, R., and A. Sharma. 2007. Antioxidant potential of mint (*Menthaspicata* L.) in radiation processed lamb meat. *Food Chem.* 100:451–8.

Kanatt, S. R., Chander, R., and A. Sharma. 2008. Chitosan glucose complex: a novel food preservative. *Food Chem.* 106:521–8.

Kim, M., and J. Han. 2002. Evaluation of physico-chemical characteristics and micro-structure of tofu containing high viscosity chitosan. *Int. J. Food Sci. Technol.* 37:277–83.

Knorr, D. 1991. Recovery and utilisation of chitin and chitosan in food processing waste management. *Food Technol.* 45:114–22.

Kubota, N., and Y. Kikuchi. 1999. Macromolecular complexes of chitosan. In *Polysaccharides structural diversity and functional versatility*, ed. S. Dumitriu, 595–628. New York: Marcel Dekker.

Lapasin, R., Stefancic, S., and F. Delben. 1996. Rheological properties of emulsions containing soluble chitosan. *Agro-Food Industry Hi-Tech* 7:12–7.

Lin, K.-W., and J.-Y. Chao. 2001. Quality characteristics of reduced-fat Chinese-style sausage as related to chitosan's molecular weight. *Meat Sci.* 59:343–51.

Maezaki, Y., Tsuji, K., Nakagawa, Y., Kawai, Y., and M. Akimoto. 1993. Hypocholesterolemic effect of chitosan in adult males. *Biosci. Biotechnol. Biochem.* 57:1439–44.

Muzzarellic, C., and A. A. A. Muzzarelli. 2006. Chitosan, a dietary supplement and a food technology commodity. In *Functional food carbohydrates*, ed. C. G. Biliaderis and M. S. Izydorczyk, Orlando, FL: Taylor and Francis.

No, H. K., Park, N. Y., Lee, S. H., Hwang, H. J., and S. P. Meyers. 2002. Antibacterial activities of chitosans and chitosan oligomers with different molecular weights on spoilage bacteria isolated from tofu. *J. Food Sci.* 67:1511–4.

Nussinovitch, A. 1997. *Hydrocolloid applications. Gum technology in the food and other industries.* London: Blackie Academic & Professional.

Nussinovitch, A. 2003. *Water soluble polymer applications in foods.* Oxford: Blackwell Publishing.

Nussinovitch, A., and M. Hirashima. 2014. *Cooking innovations, using hydrocolloids for thickening, gelling and emulsification.* Boca Raton, FL: CRC Press, Taylor & Francis Group.

Razdan, A., and D. Pettersson. 1994. Effect of chitin and chitosan on nutrient digestibility and plasma lipid concentrations in broiler chickens. *Brit. J. Nutr.* 72:277–88.

Rhee, K. S. 1987. Natural antioxidants for meat products. In *Warmed-over flavor and meat*, ed. A. J. St. Angelo and M. E. Bailey, 267–89. Orlando, FL: Academic Press.

Rudrapatnam, N. T., and S. K. Farooqahmed. 2003. Chitin—the undisputed biomolecule of great potential. *Crit. Rev. Food Sci. Nutr.* 43:61–87.

Shahidi, F., Arachchi, J. K. V., and Y. J. Jeon. 1999. Food applications of chitin and chitosans. *Trends Food Sci. Technol.* 10:37–51.

Suman, S. P., Mancini, R. A., Joseph, P., Ramanathan, R., Konda, M. K. R., Dady, G., and S. Yin. 2010. Packaging-specific influence of chitosan on color stability and lipid oxidation in refrigerated ground beef. *Meat Sci.* 86:994–8.

Terbpjevich, M., and R. A. A. Muzzarelli. 2000. Chitosan. In *Handbook of hydrocolloids*, ed. G. Phillips and P. Williams, 367–78. Cambridge: Woodhead.

Tzoumaki, M. V., Moschakis, T., Kiosseoglou, V., and C. G. Biliaderis. 2011. Oil-in-water emulsions stabilized by chitin nanocrystal particles. *Food Hydrocolloids* 25:1521–9.

Uchida, Y. 1988. Antibacterial of chitin and chitosan. *Food Chem. Japan* 2:22–9 (in Japanese).

Waliszewski, K. N., Pardio, V. T., and M. Ramirez. 2002. Effect of chitin on color during osmotic dehydration of banana slices. *Drying Technol.* 20:719–26.

Zhou, K. Q., Xia, W. S., Zhang, C., and L. Yu. 2006. In vitro binding of bile acids and triglycerides by selected chitosan preparations and their physico-chemical properties. *LWT-Food Sci. Technol.* 39:1087–92.

CHAPTER 6

Dextran

6.1 INTRODUCTION

The most important microbial polysaccharides—xanthan gum, curdlan, gellan gum, and bacterial cellulose—and their recipes were described in our previous book: *Cooking Innovations, Using Hydrocolloids for Thickening, Gelling, and Emulsification*. Additional microbial polysaccharides can have potential applications in foods, and the current chapter aims to provide the relevant details related to dextran structure, production, functionality, and regulatory status.

6.2 MANUFACTURE AND STRUCTURE

Dextran belongs to a class of glucans that are produced extracellularly by members of the genus *Lactobacillus*, *Leuconostoc*, and *Streptococcus*. It is manufactured commercially from a culture of *Leuconostoc mesenteroides* in a medium containing carbon sources—such as sucrose—and nitrogen sources—such as peptone—along with phosphates, growth factors, and trace elements. Anaerobic fermentation occurs at an initial pH of 6.7–7.2, a temperature of 25°C and in the presence of 2% sucrose.

During the initial 20 hours of fermentation, the pH drops to ~5.0 due to formation of organic acids. Dextran with increased branching is a result of elevated temperature that is maintained throughout its production process. Centrifugation is performed to remove the culture cells, followed by alcohol precipitation and purification. Dextran that is manufactured enzymatically produces a purer polymer. Better control of the enzymatic synthesis can optimize such production. Nevertheless, its cost is an issue that needs to be considered during commercial production. Dextran is a linear polysaccharide of D-glycopyranose containing α-(1,4) and α-(1,6) linkages. It is important to note that the structure of dextran differs with microbial strain and, in fact, relies on the type of dextransucrase involved. There are lower and higher (i.e., >10^4) molecular weight dextrans; the higher the molecular weight, the greater the branching of the polymer.

6.3 PROPERTIES, FOOD APPLICATIONS, AND REGULATORY STATUS

Properties of different dextrans depend on the variations in their structures. Some dextrans are water-soluble, and others are insoluble. Most of the dextrans have solubility >30 mg/mL in water. In general, dextran has high solubility and low viscosity. Dextrans can be used in foods as conditioners, stabilizers, bodying agents, and the like. Dextrans are used in confectionery products to enhance moisture retention and viscosity and inhibit sugar crystallization. In several products, such as gum and jelly candies, they can be utilized as gelling agents. Addition of dextran to ice cream will inhibit crystallization, and in some puddings, it might improve mouthfeel and sensory evaluation. Dextrans also have the potential to be used as emulsifying and thickening agents. The stability of emulsions containing sodium caseinate and dextran sulfate is quite sensitive to aqueous solution conditions. The properties of these emulsion systems can be well correlated with the corresponding complexation behavior in aqueous solutions, where either soluble or insoluble complexes are formed, depending on the pH, the ionic strength, and the protein-to-polysaccharide ratio. The stability of a mixed emulsion prepared with soluble sodium caseinate–dextran sulfate complexes is greater than that of a bilayer emulsion prepared with a two-step method, even at low dextran sulfate concentrations where bridging flocculation is observed in the bilayer emulsions. Preparation of a mixed emulsion with 0.5% sodium caseinate + 1% dextran sulfate was found to be a simple and useful way to produce a fine emulsion system, especially at a low pH, where such an emulsion cannot be prepared at all with casein(ate) alone.

Conjugation of dextran to soy protein, following dry heating of a dextran–soybean protein isolate mixture, results in improvement of the protein's physicochemical properties, especially its emulsifying and emulsion-stabilizing abilities. The stability of dressings against creaming is determined to a large extent by the state with respect to conjugation of the heated soybean protein isolate–dextran mixture, suggesting that repulsive steric forces, due to sorbed glyco-conjugate molecules, might slow down the processes, thus leading to droplet rise and cream separation. Purified whey protein isolate–dextran conjugate produced in aqueous solution under mild conditions showed good solubility over the pH range of 3.2–7.5 at high ionic strengths of 0.05–0.2 M. It also showed high heat stability compared to native whey protein isolate. The emulsifying ability and stability of emulsions made with the whey protein isolate–dextran conjugate were greatly improved compared with the native whey protein isolate or a natural commercial glycoprotein emulsifier, such as gum arabic. At some point, whey protein isolate–dextran conjugates can become an alternative to gum arabic as a food ingredient in emulsions. The improved heat and pH stability of protein conjugates suggest that these ingredients can help in fortifying acid beverages that are heat-processed. Another study dealt with conjugation of carp myofibrillar proteins with dextran (Mf–Dex) through the Maillard reaction. Lyophilized myofibrils were mixed with dextran (in a weight ratio of 1:9). The mixture was kept at two different temperatures, 40°C and 50°C (65% relative humidity), to prepare neoglycoprotein. The protective effect of dextran was required to prepare Mf–Dex with high solubility, and the myosin-heavy chain was selectively conjugated with dextran. Mf–Dex developed an excellent emulsifying property, and its solubility remained almost unchanged by heating at 50°C for 6 hours. The improved emulsion stability was not impaired by heat treatment at 50°C. Dextrans do not have clear approval as a food additive in the United States or Europe. Nevertheless, *L. mesenteroides* is generally regarded as safe (GRAS), and its presence in various fermented foods has been permitted. Moreover, several patents demonstrate its ability to be used in baked goods.

6.4 RECIPES

6.4.1 Dinner Rolls (Figure 6.1)

(Makes 12 rolls)

7.5 g (a little under 2 tsp) dextran
250 g (1⅘ cups) strong wheat flour
180 mL (a little under ¾ cup) lukewarm water (38°C)
7 g (2 tsp) dry yeast
18 g (2 tbsp) white sugar
5 g (a little under 1 tsp) salt
1 egg, beaten
40 g unsalted butter
25 mL (a little under ⅓ cup) skimmed milk

1. Soften unsalted butter and beat egg.
2. Put dextran, wheat flour, lukewarm water, dry yeast, white sugar, salt, softened butter, ½ of the beaten egg (1) (25 g), and skimmed milk in a bread machine, and follow the machine's instructions for the first rising (Figure 6.1A) (Hint 6.4.1.1).
3. Punch down and divide the dough (2) into 12 pieces (about 45–50 g) with a dough scraper, shape into rounds, and let them stand for 15 minutes covered with a damp dish towel (Figure 6.1B). Afterwards cover with a dry towel (Figure 6.1C).
4. Form each dough ball (3) into a cone shape, and let it stand for another 10 minutes covered with a damp dish towel and dry towel (Figure 6.1D).
5. Roll each cone of the dough (4) out into a triangle with a rolling pin (Figure 6.1E), while pulling its edges, then roll it up (Figure 6.1F).

FIGURE 6.1 DINNER ROLLS. (A) LET THE DOUGH RISE IN THE BREAD MACHINE. (B) DIVIDE THE DOUGH AND COVER WITH A DAMP DISH TOWEL. (C) COVER WITH DRY TOWEL. (D) SHAPE DOUGH INTO CONES; LET IT STAND FOR 10 MINUTES. (E) ROLL OUT WITH A ROLLING PIN. (F) ROLL THE DOUGH TO SHAPE IT. (G) PLACE DOUGH ON BAKING PAPER AND SPRINKLE WITH WATER. (H) LET DOUGH RISE AND BRUSH WITH EGG WASH. (I) BAKE ROLLS.

6. Spread baking paper on a baking tray, place the dough (5) on it with the rolled edge facing down, and sprinkle water on it (Figure 6.1G).
7. Let the dough (6) rise again at 30°C–35°C for 40 minutes.
8. Preheat the oven to 200°C, spread the remaining beaten egg on the fermented dough (7) (Figure 6.1H), and bake for 15 minutes (Figure 6.1I).

pH of dough = 5.51

Preparation hints:

6.4.1.1 To ferment the dough without using a bread machine, see Chapter 14.

Adding dextran to this recipe results in a softer dough with greater volume and longer shelf life.

6.4.2 Ice Cream Cones (Figure 6.2)

(Makes 5 cones)

20 g (2⅔ tbsp) dextran
80 g (⅗ cup) all-purpose flour
30 g (1⅔ tbsp) maple syrup
10 mL (2 tbsp) skimmed milk
55 mL (3⅔ tbsp) water
20 g butter
Thick paper

1. Cut thick paper (Hint 6.4.2.1) into five squares (about 14 cm × 14 cm), form into cones, and wrap baking paper around each (Figure 6.2A).
2. Melt butter.
3. Mix dextran, all-purpose flour, and skimmed milk in a bowl, add the melted butter (2), maple syrup, and water (Figure 6.2B), and knead.
4. Divide the dough (3) into five pieces and form into balls.
5. Roll each dough ball (4) out into a round (about 13 cm in diameter) with a rolling pin (Figure 6.2C), and wrap around formed paper (Figure 6.2D).

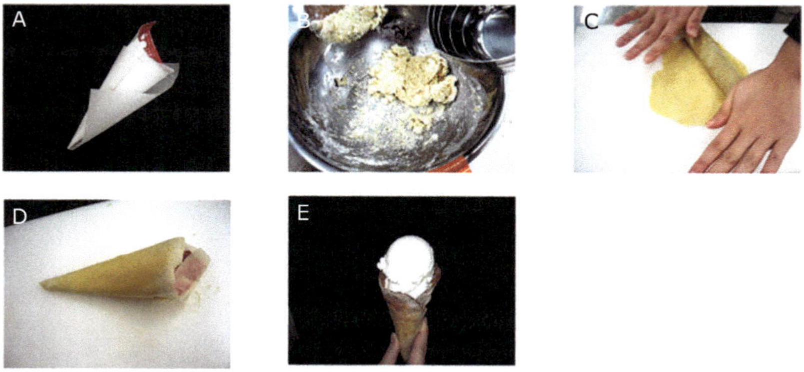

Figure 6.2 Ice cream cones. (A) Form carton pieces into cones and wrap with baking paper. (B) Mix dextran, flour, and skimmed milk and add melted butter, maple syrup, and water. (C) Roll the dough out. (D) Wrap rolled dough around paper cone to form shape. (E) Serve baked cones with ice cream.

6. Preheat oven to 180°C, bake for 10–15 minutes, and cool.
7. Remove the paper, and fill cones with ice cream (Figure 6.2E).

pH of dough = 5.68

Preparation hints:
6.4.2.1 A milk carton or juice carton can be used.

This recipe has a crispy, frangible, light texture. The cones can be stacked without wedging or sticking.

6.5 TIPS FOR THE AMATEUR COOK AND PROFESSIONAL CHEF

- Incorporation of small quantities of dextran in yeast-raised bread doughs containing both yeast and gluten produces softer baked goods with greater volume and longer shelf life.
- Addition of dextran to dough increases its water-absorptive properties and makes it more extensible.
- Dextran can stabilize chocolate drink beverages as well as soft drinks and flavor extracts.

REFERENCES AND FURTHER READING

Day, D. F., and D. Kim. 1993. Process for the production of dextran polymers of controlled molecular size and molecular size distributions. US Patent 5,229,277.

Diftis, N. G., Biliaderis, C. G., and V. D. Kiosseoglou. 2005. Rheological properties and stability of model salad dressing emulsions prepared with a dry-heated soybean protein isolate–dextran mixture. *Food Hydrocolloids* 19:1025–31.

European Commission. 2000. Opinion of the scientific committee on food on a dextran preparation, produced using *Leconostoc mesenteroides*, *Saccharomyces cerevisa* and *Lacobacillus* spp. as a novel food ingredient in bakery products. Document CS/NF/DOS/7/ADD 3 FINAL.

Fujiwara, K., Oosawa, T., and H. Saeki. 1998. Improved thermal stability and emulsifying properties of carp myofibrillar proteins by conjugation with dextran. *J. Agric. Food Chem.* 46:1257–61.

Gounga, M. E., Xu, S.-Y., Wang, Z., and W. G. Yang. 2008. Effect of whey protein isolate–pullulan edible coatings on the quality and shelf life of freshly roasted and freeze-dried Chinese chestnut. *J. Food Sci.* 73:E155–61.

Jourdain, L., Leser, M. E., Schmitt, C., Michel, M., and E. Dickinson. 2008. Stability of emulsions containing sodium caseinate and dextran sulfate: Relationship to complexation in solution. *Food Hydrocolloids* 22:647–59.

Khan, T., Park, J. K., and J.-H. Kwon. 2007. Functional biopolymers produced by biochemical technology considering applications in food engineering. *Korean J. Chem. Eng.* 24:816–26.

Leathers, T. D., Hayman, G. T., and G. L. Cote. 1997. Microorganism strains that produce a high proportion of alternan to dextran. US patent 5,702,942.

Nussinovitch, A. 1997. *Hydrocolloid applications. Gum technology in the food and other industries.* London: Blackie Academic & Professional.

Nussinovitch, A. 2003. *Water soluble polymer applications in foods.* Oxford: Blackwell Publishing.

Nussinovitch, A. 2010. *Plant gum exudates of the world sources, distribution, properties and applications.* Boca Raton, FL: CRC Press, Taylor & Francis Group.

Nussinovitch, A., and M. Hirashima. 2014. *Cooking innovations, using hydrocolloids for thickening, gelling and emulsification.* Boca Raton, FL: CRC Press, Taylor & Francis Group.

Park, J. K., and T. Khan. 2009. Other microbial polysaccharides: Pullulan, scleroglucan, elsinan, levan, alternan, dextran. In *Handbook of hydrocolloids*, ed. G. O. Phillips and P. A. Williams, 592–615. Boca Raton, FL: CRC and Oxford: Woodhead Publishing Limited.

Zhu, D., Damodaran, S., and J. A. Lucey. 2010. Physicochemical and emulsifying properties of whey protein isolate (WPI)–dextran conjugates produced in aqueous solution. *J. Agric. Food Chem.* 58:2988–94.

CHAPTER 7

Gum Ghatti

7.1 HISTORICAL BACKGROUND

Since ancient times, gum ghatti has had a reputation in India as a fabulous medicinal product, in addition to having recognized qualities for food items. The gum is mentioned in very old therapeutic scriptures such as the *Ayurveda* (Indian system of medicine). In a number of regions in India, the clean nodules of a certain variety of gum ghatti are consumed as a comfort food and are regarded as a sign of prosperity.

7.2 COMMON NAMES, DISTRIBUTIONAL RANGE, AND ECONOMIC IMPORTANCE

Gum ghatti is an amorphous, translucent exudate of the *Anogeissus latifolia* tree of the family *Combretaceae*. Gum ghatti is also known as button tree, *dindiga* tree, *ghatti* tree, *baklee* [India], and *dhaura* [India]. The term "ghatti gum" is generally used in European commerce for any highly viscous gum of Indian origin. In the old days, after being collected and sun-dried, the gum was transported to Bombay by land through mountain ghats or passes, hence the name ghatti. The native distributional range of gum ghatti covers India, Nepal, Pakistan, and Sri Lanka. Gum ghatti is used sparsely in food, drug, and cosmetic products, mainly as a stabilizer for oil-in-water emulsions. It has not found a large variety of applications in foods, perhaps due to the variably quality of the commercially offered gum but also because of the relatively small quantity of the gum that is accessible to the world market. Other parts of the tree also have commercial and functional uses, for example, the wood can be used in axle handles, construction, agricultural implements, house posts, poles, fuel, and charcoal. Similarly, the leaves are used for tanning, and the bark is useful in medicinal preparations.

7.3 EXUDATE APPEARANCE

Gum ghatti and gum karaya (Chapter 8) are found in very similar geographical regions, and consequently, the harvesting and grading techniques for the two are also similar. The gum is collected on a regular basis in April from tapped trees. It is produced as formless or rounded tears (0.5–1.0 cm in diameter) or as larger vermiform pieces with glassy fractures. The gum has an insignificant smell and a non-distinctive taste. Gum left on the tree through the monsoon season turns into a dark, agglutinated mass. The average seasonal yield of tapped gum per tree is 60–100 g, with a larger exudation yield in dry years. In *A. latifolia*, there is no natural preformed gum-producing tissue system in either the bark or the wood. Natural wounds, for instance the breaking of branches by wind, cause exudation. Again, heavy tapping damages the cambium and makes wound healing difficult, thus limiting the lifespan of the tree. In India, an ~450-fold boost in gum yield was recorded in trees treated with 1,600 mg of ethephon during April–May, when the trees are leafless.

7.4 EXUDATE COLOR AND SOLUBILITY

The highest grade is pale tan and almost completely free of adhering bark. The inferior grades fluctuate from medium tan to dark brown and may include as much as 7% insoluble impurities. Powdered gum has a gray to reddish-gray color. The gum of *A. latifolia* does not dissolve in water to yield a clear solution, but rather over 90% of the gum forms a colloidal dispersion. The gum may be rendered soluble by autoclaving.

7.5 CHEMICAL CHARACTERISTICS

Structural and chemical characteristics of *A. latifolia* have been studied. One of the major structural features of *A. latifolia* gum is its linear arrangement of (1,6)-linked D-galactopyranoses. It is composed of L-arabinose, D-galactose, D-mannose, D-xylose, and D-glucuronic acid in a molar ratio of 10:6:2:1:2 with trace quantities (less than 1%) of 6-deoxyhexose. Upon graded hydrolysis, two aldobiuronic acids, 6-O-(β-D-glucopyranosyluronic acid)-D-galactose and 2-O-(β-D-glucopyranosyluronic acid)-D-mannose, are obtained. In addition, 50% pentose and 12% galactose or galacturonic acid are present. Partial acid hydrolysis yields two homologous series of oligosaccharides together with small amounts of 3-O-β-D-galactopyranosyl-D-galactose and 2-O-(β-D-glucopyranosyluronic acid)-D-mannose. Acid-labile side chains are attached to the backbone through L-arabinofuranose residues. The soluble portion of the gum, which contains 0.72% nitrogen, consists of calcium, magnesium, potassium, and sodium salts of a polysaccharide acid. With respect to its constituent sugars, gum ghatti is highly reminiscent of acacia, damson, cherry, egg plum, and mesquite gums (Chapter 13), all of which include a high proportion of terminal L-arabinofuranose units. One of the aldobiuronic acids obtained upon hydrolysis, i.e., 6-O-β-D-glucopyranosyluronic acid-D-galactose, is also found in gum arabic. But gum ghatti differs from the above gums in its (1,6)-linked galactose framework. It includes 10%–12% moisture and up to 3% acid-insoluble ash. When gum ghatti of a particular grade is spray dried, all the insoluble matter in it is removed and the resultant product has much lower viscosity owing to processing conditions and hydrolysis. The gum is not colored blue by iodine solution, indicating a lack of starch and dextrin. It is insoluble in alcohol.

7.6 PHYSICAL PROPERTIES

The molecular weight of the gum's soluble portion is about 12,000. Small amounts of acid or alkali do not have any effect on gum ghatti dispersions, given that the gum acts as a buffer and reverts to its normal pH (about 5.5). Nevertheless, large quantities of acid or alkali will overcome this buffering action. Gum ghatti solutions lose their viscosity at high pH. When dispersed in water at high concentrations (over 10%), gum ghatti forms a viscous, adhesive mucilage, which is less viscous than that of gum karaya (Chapter 8), but more viscous than that of gum arabic. Its adhesive properties are not as strong as those of gum arabic. Other rheological properties of *A. latifolia* solutions are reviewed elsewhere.

7.7 COMMERCIAL AVAILABILITY OF THE GUM AND APPLICATIONS

Although Indian ghatti is for the most part obtained from *A. latifolia*, there is no doubt that gums from absolutely dissimilar botanical sources are often referred to as gum ghatti by Indian merchants and are exported under that name. Such substitutes are collected from *Terminalia alata*, *Terminalia bellerica*, *Terminalia tomentosa*, *Bauhinia variegata*, *Acacia catechu*, and *Acacia arabica* (Indian gum arabic). Gum ghatti should not be confused with either "Bassora gum" (gums of this class somewhat resemble gum tragacanth in their gelling properties, but are darker) or *Sterculia* gum, each of which is sometimes referred to as "Indian gum".

The uses of gum ghatti are comparable to those of gum arabic: in foods, for tablet binding, and for emulsification purposes. It was once used extensively in oil-well drilling fluids to enhance their viscosity and thixotropy, and to minimize fluid loss. Gum ghatti can be used as an emulsifier in products that gum arabic fails to emulsify. "Gatifolia", a new gum ghatti product, is manufactured by a non-chemical physical procedure that involves dissolution, filtering, sterilization, and spray-drying. It is claimed to have better emulsification ability, acid resistance, and salt tolerance. Gatifolia has been reported to offer an additional, extensively distributed range of proteinaceous molecular components to bind oil, which makes it superior to other natural emulsifiers. The gum has low viscosity, but it is more soluble in water than many other polysaccharides. In oriental foods, such as instant rice noodles, gum ghatti and fenugreek gum are utilized to improve textural attributes and mouthfeel. Decreased oil content of deep-fat-fried foods would be welcomed by both manufacturers and consumers, and the inclusion of hydrocolloid additives is believed to be most promising in this respect. Gum arabic and carrageenan were observed to be most effective when they were added to a product prepared from a soft chickpea flour dough that is prepared by adding 45 mL water to 100 g flour. Gum ghatti was less effective, perhaps due to its water-retention capacity. Gum ghatti is known to form films that are soluble and brittle and are not thought to be very useful. However, the type of film that might result from possible interactions of gum ghatti in fried foods is not known. Other hydrocolloids, such as xanthan, are used today as substitutes for gum ghatti. Powdered gum ghatti has also been used in explosives to keep the ammonium nitrate dry in wet ground. The gum absorbs any water which seeps into the explosive cartridge and swells to form an insulating surface layer. It has also been employed in paints as a glaze or varnish, as a binder in coating compositions, and in ceramics as a binder to enhance the wet strength of clay prior to firing.

7.8 RECIPES

7.8.1 Mayonnaise-Type Dressing (Figure 7.1)

(Makes 180 g)

4 g (a little under 2 tsp) gum ghatti powder

0.3 g (1/6 tsp) xanthan gum powder

2 g (2/3 tsp) white sugar

5 g (a little under 1 tsp) salt

20 mL (1 1/3 tbsp) water

FIGURE 7.1 MAYONNAISE-TYPE DRESSING. (A) MIX TOGETHER GUM GHATTI, XANTHAN GUM, SUGAR, SALT, AND MONOSODIUM GLUTAMATE WITH GRADUAL ADDITION OF WATER. (B) STIR VINEGAR INTO THE GUM GHATTI SOLUTION. (C) STIR VEGETABLE OIL INTO THE GUM GHATTI SOLUTION, EMULSIFY, AND SEASON.

140 g (a little over ⅔ cup) vegetable oil*
25 mL (1⅔ tbsp) vinegar*
Monosodium glutamate, ground black pepper

1. Mix gum ghatti, xanthan gum, sugar, salt, and monosodium glutamate in a bowl, and blend in water gradually (Figure 7.1A).
2. Stir vinegar into the gum ghatti solution (1) (Figure 7.1B).
3. Stir vegetable oil gradually into the gum ghatti solution (2) (Hint 7.8.1.1) until it is completely emulsified, and season with pepper or other spices (Figure 7.1C).

pH=3.25

Preparation hints:

7.8.1.1 It is important that the vegetable oil be added gradually for complete emulsification.

The mayonnaise has a very thick texture, despite the absence of eggs. This recipe can, therefore, be used for people with egg allergies.

7.8.2 TABLE SYRUP (FIGURE 7.2)

(Makes 180 g)

150 g (a little over ⅔ cup) cane syrup

* You can use your favorite oil and vinegar.

0. 8 g (a little over ⅓ tsp) gum ghatti powder
30 g (1⅓ tbsp) corn syrup
10 g (½ tbsp) maple syrup
0.04 g citric acid, anhydrous
0.12 g trisodium citrate dehydrate
4 g (1 tsp) butter
0.2 g lecithin

1. Heat 120 g (a little under ⅓ cup) cane syrup to 95°C in a bottle in a water bath, add gum ghatti (Figure 7.2A) (Hint 7.8.2.1), and hold the temperature at 80°C with stirring using a magnetic stirrer for 60 minutes (Figure 7.2B, left).
2. Mix the remaining cane syrup, corn syrup, maple syrup, citric acid, and trisodium citrate dehydrate in another bottle, and heat it at 80°C with stirring using a magnetic stirrer for 5 minutes (Figure 7.2B, right).
3. Add the syrup mixture (2) to the cane syrup and gum ghatti (1) (Figure 7.2C), mix for a while with a magnetic stirrer, and pour into a bowl.

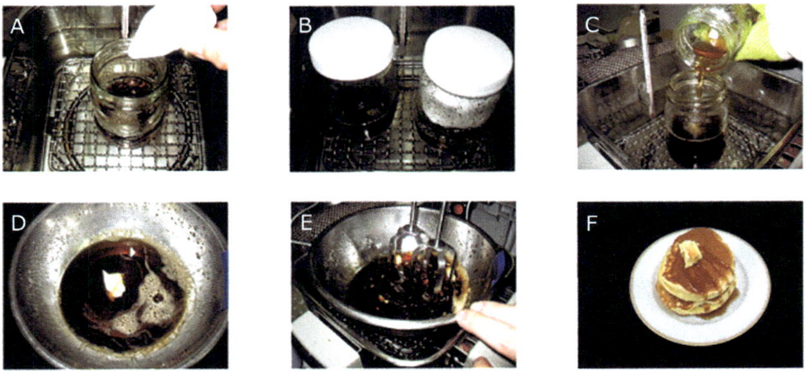

Figure 7.2 Table syrup. (A) Heat cane syrup and add gum ghatti. (B) Hold the temperature at 80°C while stirring. (B, left) Cane syrup and gum ghatti mixture. (B, right) Cane syrup, corn syrup, maple syrup, citric acid, and trisodium citrate dehydrate mixture. (C) Combine both mixtures and stir. (D) Add butter and lecithin. (E) Stir. (F) Serving suggestion: use syrup on pancakes.

4. Add butter and lecithin to the mixture (Figure 7.2D), and stir well with a blender at 80°C for 5 minutes (Figure 7.2E).

pH=4.32

Preparation hints:

7.8.2.1 It is hard to dissolve gum ghatti in cane syrup because there is little water. You need to stir vigorously to produce a solution without lumps.

This recipe is good for pancakes (Figure 7.2F). It was prepared in the laboratory using an experimental apparatus, but the temperature and stirring can be controlled with regular cooking utensils.

7.9 TIPS FOR THE AMATEUR COOK AND PROFESSIONAL CHEF

- The viscosity of gum ghatti increases exponentially with increasing concentration.
- At relatively dilute concentrations (5% gum ghatti and 10% gum arabic), both gums show evidence of shear-thinning behavior at low shear rates and Newtonian behavior at higher shear rates.
- Gum ghatti can emulsify 20% medium-chain triglyceride oil at a much lower concentration (5 wt%) than gum arabic.

7.10 REGULATORY STATUS

Gum ghatti is permitted in the United States as an emulsifier in non-alcoholic beverages and beverage bases at a maximum level of 0.2%. It is permitted in all other food categories at a maximum level of 0.1%. In Japan, gum ghatti is permitted in all food items as a natural food additive with no specific function defined.

REFERENCES AND FURTHER READING

Al-Assaf, S., Amar, V., and G. O. Phillips. 2008. Characterization of gum ghatti and comparison with gum Arabic. In *Gums and stabilizers for the food industry 14*, ed. P. A. Williams and G. O. Phillips, 280–90. Wrexham: Royal Society of Chemistry.

Annapure, U. S., Singhal, R. S., and P. R. Kulkarni. 1999. Screening of hydrocolloids for reduction in oil uptake of a model deep fat fried product. *Fett/Lipid* 101:217–21.

Aspinall, G. O., Auret, B. J., and E. L. Hirst. 1958. Gum ghatti (Indian gum). Part II. The hydrolysis products obtained from the methylated degraded gum and the methylated gum. *J. Chem. Soc.* 221–30.

Aspinall, G. O., Auret, B. J., and E. L. Hirst. 1958. Gum ghatti (Indian gum). Part III. Neutral oligosaccharides formed on partial acid hydrolysis of the gum. *J. Chem. Soc.* 4408–14.

Aspinall, G. O., Bhavanandan, B. V., and T. B. Christensen. 1965. Gum ghatti (Indian gum). Part V. Degradation of the periodate-oxidised gum. *J. Chem. Soc.* 2677–84.

Aspinall, G. O., and T. B. Christensen. 1965. Gum ghatti (Indian gum). Part IV. Acidic oligosaccharides from the gum. *J. Chem. Soc.* 2673–6.

Aspinall, G. O., Hirst, E. L., and A. Wickstrom. 1955. Gum ghatti (Indian gum). The composition of the gum and the structure of two aldobiuronic acids derived from it. *J. Chem. Soc.* 1160–5.

Elworthy, P. H., and T. M. George. 1963. The molecular properties of ghatti gum. A naturally occurring polyelectrolyte. *J. Pharm. Pharmacol.* 15:781–93.

Fleischer, J. 1959. Gum ghatti. In *Industrial gums*, ed. R. L. Whistler, 311–20. New York: Academic Press.

Howes, F. N. 1949. *Vegetable gums and resins.* Waltham, MA: Chronica Botanica Company.

Ido, T., Ogasawara, T., Katayama, T., Sasaki, Y., Al-Assaf, S., and G. O. Phillips. 2008. Emulsification properties of Gatifolia (Gum Ghatti) used for emulsions in foodproducts. *Foods Food Ingredients J. Jpn.* 213:365–71.

Jefferies, M. J., Pass, G., and G. O. Philips. 1977. Viscosity of aqueous solutions of ghatti. *J. Sci. Food Agric.* 28:173–9.

Koh, L., Jiang, B., Kasapis, S., and C. Woo Foo. 2011. Structure, sensory and nutritional aspects of soluble-fibre inclusion in processed food products. *Food Hydrocolloids* 25:159–64.

Meer, G. 1980. Gum ghatti. In *Handbook of water-soluble gums and resins*, ed. R. L. Davidson, 9. 1–9. 8. New York: McGraw-Hill.

Nussinovitch, A. 1997. *Hydrocolloid applications. Gum technology in the food and other industries*. London: Blackie Academic & Professional.

Nussinovitch, A. 2003. *Water soluble polymer applications in foods*. Oxford: Blackwell Publishing.

Nussinovitch, A. 2010. *Plant gum exudates of the world sources, distribution, properties and applications*. Boca Raton, FL: CRC Press,Taylor & Francis Group.

Nussinovitch, A., and M. Hirashima. 2014. *Cooking innovations, using hydrocolloids for thickening, gelling and emulsification*. Boca Raton, FL: CRC Press, Taylor & Francis Group.

Pszczola, D. E., and K. Banasiak. 2006. Enter IFT's magic ingredient kingdom. *Food Technol*. 60:45–92.

CHAPTER 8

Gum Karaya

8.1 INTRODUCTION

Exudate gums have been employed for centuries in many fields; they have maintained their significance despite the arrival of a variety of new gums with comparable and distinctive performances. The gums exude from trees and shrubs in the form of tear-like, striated nodules or as amorphous lumps, and then dry in the sun, forming hard, glassy, different-colored exudates. Gum production increases under elevated temperatures and limited moisture. Yields can be enhanced by making incisions in the bark or by stripping it from the tree or shrub. Exudate gums have been exploited in food applications for years, for the purpose of emulsification, thickening, and stabilization. Tragacanth and karaya gums are considered safe for human consumption based on a long and non-toxic history of use; more recent toxicological studies also lead to the same conclusions. Tree gum exudates can also be utilized in non-food applications, such as pharmaceuticals, cosmetics, textiles, lithography, and minor forest-manufactured goods.

8.2 GEOGRAPHICAL DISTRIBUTION

The genus *Sterculia* includes over a hundred species that are distributed across the warmer parts of the world. The *S. urens* trees are generally found in dry, rocky hills and plateaus and are widespread in the dry, deciduous forests of northern and central India. It also occurs on the west coast of India, and in the dry forests of Myanmar and Sri Lanka.

8.3 EXUDATE APPEARANCE

S. urens gum exudes naturally; however, nearly all of it is manufactured by artificial stimulation, through a particular tapping method. In this process, segments of the trunk are blazed or the bark is cut away. Gum immediately begins to exude and this continues for a few days; however, exudation is maximum during the first 24 hours after blazing. The crude gum is set aside to dry on the tree, and once it is removed, supplementary gum formation is encouraged by scraping the wound to expose fresh surface. The tapping or collection season is approximately from October–January and from April–June. As the weather becomes warmer, gum yields increase. Early rains decrease the mass of the harvest by washing away a great deal of the exudate faster than it can dehydrate. The common tree can be tapped about five times throughout its lifetime, with 1–5 kg exudate formed per tree per season. The gathered gum is broken into smaller, unevenly shaped pieces with a mallet, and they have a somewhat crystalline form or they might look like broken glassy tears. A number of regions in India have both red- and white-barked trees. The former is thought to produce more gum, and it is also thought that the trees on hill slopes provide more gum than those in other locations. Ethephon was used to increase karaya gum production from *S. urens*. The trees treated with ethephon yielded about 20 times more gum than the non-treated trees: the entire harvest from seven treated trees, each tapped once, was about 1.5 kg of high-quality gum. The maximum gum-exudation response was obtained with 768 mg of active ethephon. Histological and histochemical changes during the development of gum canals in *S. urens* have been studied. The highest grades of *S. urens* gum are white and translucent and are nearly free of bark. The inferior grades are colored light yellow to brown. In its ground form, the gum is white to grayish-white.

8.4 WATER SOLUBILITY

All *Sterculia* gums swell in water. The gum's powder size influences the type of water dispersion. A coarse granulation (6–30 mesh) of *S. urens* gum gives a discontinuous grainy dispersion. The gum swells to 60–100 times its starting volume; a delicately powdered gum (150–200 mesh) yields a homogeneous dispersion. Gum karaya contains about 40% uronic acid residues and up to ~8% acetyl groups. It is because of these substituents that the gum does not dissolve entirely in water, and instead swells. By the means of chemical deacetylation, the gum can be modified from a water-swellable to a water-soluble material. Gum karaya contains the lowest amount of proteinaceous material among all the exudate gums. Following dispersion in water, the gum absorbs the water to form a thick solution, and yield stresses of 60 and 100 mN/cm^2 have been found for 2% and 3% gum concentrations, respectively. The smoothness of the gum solution is determined by its particle size, but extended stirring can produce a smooth texture and lower the viscosity. Gum solubility can be increased by deacetylation, which gives the product a more expanded conformation. The solutions are cohesive and stringy or ropy. Heating changes the polymer conformation and increases the solubility. Ropiness is accompanied by lower acetyl content. Because heating increases solubility and the reduction in viscosity is irreversible, solution concentrations can be increased to 15%.

8.5 CHEMICAL CHARACTERISTICS

S. urens gum occurs in nature as a complex, to some extent acetylated, branched polysaccharide with a molecular weight of about 9,500,000. It has about 40% uronic acid residues and roughly 8% acetyl groups. The gum is a calcium and magnesium salt, with a central chain of D-galactose (13%), L-rhamnose (15%), and D-galacturonic acid (43%) units, with a number of side chains containing D-glucuronic acid.

Powdered gum karaya has approximately 14%–18% moisture, less than 1% acid-insoluble ash, and less than 3% insoluble substance or bark. It has a tendency to develop acetic acid upon contact with humid air. Indian gum karaya is unlike the African variety in its elevated acid value and in its more pronounced acetic acid odor. The pH of a 1% gum karaya (of Indian origin) solution is 4.4–4.7 and it is 4.7–5.2 for the African gum karaya. When pH > 7.0, alkali irreversibly alters the distinctive short-bodied gum karaya that is swollen in solution into a ropy mucilage. This is due to the deacetylation of the gum karaya molecules. The molecular structure and other chemical features of *S. urens* gum have been described in numerous studies.

8.6 COMMERCIAL AVAILABILITY AND FOOD APPLICATIONS

At the beginning of the 20th century, gum karaya was sold as gum tragacanth or as an adulterant of gum tragacanth, due to the large price gap between the two gums. Nevertheless, gum karaya was proven to have better applications than accessible gums and, thus, it found its own niche in the marketplace. The gum is employed globally in a variety of applications, although some of these applications are now using more recent synthetic, lower-cost gums. It is utilized as a stabilizing agent in food, such as water ice, soft candy, meringues, and cheese spreads.

8.7 RECIPES

8.7.1 Fruit Sherbet (Figure 8.1)

(Serves 5)

120 g (a little over ½ cup) granulated sugar

2 g (a little under 1 tsp) gum karaya powder

2 g (⅔ tsp) guar gum powder

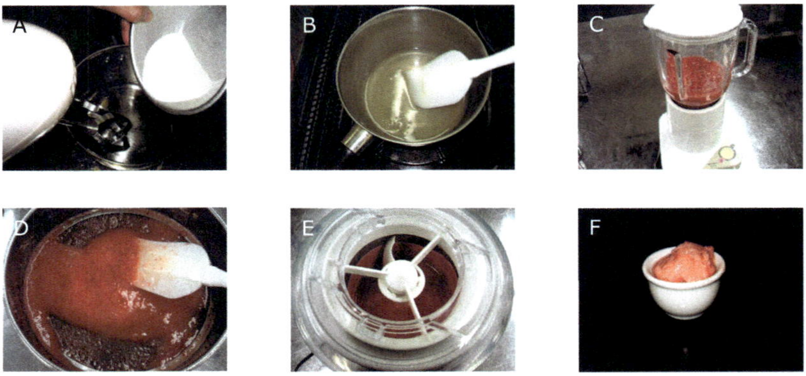

Figure 8.1 Fruit sherbet. (A) Mix granulated sugar, gum karaya, and guar gum and add to water while stirring. (B) Heat the dispersion to dissolution, boil, and cool. (C) Purée fresh fruit. (D) Strain the fruit purée. (E) Put the sweetened fruit purée dispersion in an ice cream maker. (F) Serve strawberry sherbet.

240 mL (a little under 1 cup) water
800 g fresh fruit
100 mL (⅖ cup) lime or lemon juice

1. Mix granulated sugar, gum karaya, and guar gum in a bowl and add to water while stirring with a hand mixer (Figure 8.1A).
2. Heat the dispersion (1) with stirring, until it dissolves at low heat (at 85°C), and then boil it for 1 minute.
3. Cool the solution (2) to less than 30°C (Figure 8.1B).
4. Purée fresh fruit in a blender (Figure 8.1C) and strain (Figure 8.1D).
5. Add lime juice to the strained fruit (4).
6. Put the cooled solution (3) and fruit purée (5) into an ice cream maker (Figure 8.1E), follow the manufacturer's instructions, and serve (Figure 8.1F).

pH of strawberry sherbet = 3.14

This recipe produces a smooth-textured sherbet. Remove it from the freezer 10–20 minutes before eating.

8.8 TIPS FOR THE AMATEUR COOK AND PROFESSIONAL CHEF

- Gum karaya absorbs water quickly to produce viscous colloidal sols.
- Heating dispersions of gum karaya results in permanently lower viscosities, but increased gum solubility.
- Addition of 0.2%–0.4% gum karaya prevents the development of large ice crystals in ice pops and sherbets.
- Using gum karaya at ~0.25% in ground meat products gives them a smooth appearance.
- Karaya, in combination with carrageenan or alginates, retards staling of baked goods, including bread.
- Karaya, as well as locust bean gum and carrageenan, are effective in stabilizing whipped-cream products.

REFERENCES AND FURTHER READING

Anderson, D. M. W. 1989. Evidence for the safety of gum karaya (*Sterculia* spp.) as a food additive. *Food Additives and Contaminants* 6:189–99.

Aspinall, G. O., Khondo, L., and B. A. Williams. 1987. The hex-5-enose degradation: Cleavage of glycosiduronic acid linkages in modified methylated Sterculia gums. *Can. J. Chem.* 65:2069–76.

Aspinall, G. O., Fanous, H. K., Kumar, N. S., and V. Puvanesarajha. 1981. The selective cleavage of permethylated glycopyranosiduronic acid linkages by oxidative decarboxilation with lead tetraacetate. *Can. J. Chem.* 59:935–40.

Aspinall, G. O., and U. D. Nasir. 1965. Plant gums of the genus Sterculia. Part I. The main structural features of *Sterculia urens* gum. *J. Chem. Soc.* 2710–20.

Aspinall, G. O., and V. Puvanesarajah. 1984. The hex-5-enose degradation: cleavage of 6-deoxy-6-iodo-α-D-galactopyranoseidic linkages in modified carboxyl-reduced methylated tragacanthic acid. *Can. J. Chem.* 62:2736–9.

Aspinall, G. O., and G. R. Sanderson. 1970. Plant gums of the genus Sterculia. IV. Oligosaccharidesfrom acidic *Sterculia urens* gum. *J. Chem. Soc.* 16:2256–8.

Aspinall, G. O., and G. R. Sanderson. 1970. Plant gums of the genus *Sterculia.* V. Degradation of carboxy-reduced *Sterculia urens* gum. *J. Chem. Soc.* 16:2259–64.

Babu, A. M., and A. R. S. Menon. 1989. Ethephon induced gummosis in *Bombax ceiba* L. and *Sterculia urens* Roxb. *Indian Forester* 115:44–7.

Balls, E. K. 1962. *Early uses of California plants.* Berkeley: The University of California Press.

Ben-Zion, O., and A. Nussinovitch. 1997. Physical properties of hydrocolloid wet glues. *Food Hydrocolloids* 11:429–42.

Edwards, H. G. M., Falk, M. J., Sibley, M. G., Alvarez, J. B., and F. Rull. 1998. FT-Raman spectroscopy of gums of technological significance. *Spectrochimica Acta, Part A—Mol. Biomol. Spectroscopy* 54A:903–20.

Felker, P., and R. S. Bandurski. 1979. Uses and potential uses of leguminous trees for minimal energy input agriculture. *Econ. Bot.* 33:172–84.

Felter, H. W., and J. U. Lloyd. 1898. *King's American dispensatory*, 18th edn., 3rd revision, reprinted 1983. Portland, OR: Eclectic Medical Publications.

Fowells, H. A. 1965. U.S. Department of Agriculture, Forest Service, Silvis of forest trees of the United States. Washington DC: Agriculture Handbook 271.

Glicksman, M. 1963. Natural gums. In *Kirk-Othmer encyclopedia of chemical technology*, 2nd edn., vol. 10, 741–54. New York: Wiley.

Glicksman, M. 1983. Gum karaya (*Sterculia gum*) [Manufacture, structure, properties, applications in pharmaceuticals and foods, regulatory status]. *Food Hydrocolloids* 2:39–47.

Glicksman, M. 1983. Gum tragacanth: Regulatory status, structure, properties, food applications (*Astragalus gummifer*). *Food Hydrocolloids* 2:49–60.

Goldstein, A. M., and E. N. Alter. 1959. Gum karaya. In *Industrial gums*, ed. R. L. Whistler, 343–59. New York: Academic Press.

Howes, F. N. 1949. *Vegetable gums and resins.* Waltham, MA: Chronica Botanica Company.

Imeson, A. P. 1992. Natural plant exudates. In *Thickening and gelling agents for food*, ed. A. P. Imeson. London: Blackie Academic & Professional.

Kesar, S. 1935. Kullu (*Sterculia urens*). Wailings from Damoh, C. P. *Indian Forester* 61:144–9.

Kubal, J. V., and N. Gralen. 1948. Physico-chemical properties of karaya gum. *J. Colloid Sci.* 3:457–71.

Lewis, W. H., and M. P. F. Elvin-Lewis. 1977. *Medical botany.* New York: John Wiley & Sons.

Mantell, C. L. 1947. *The water-soluble gums.* New York: Reinhold Publ. Corp.

Mills, P. L., and J. L. Kokini. 1984. Comparison of steady shear and dynamic viscoelastic properties of guar and karaya gums. *J. Food Sci.* 49:1–4, 9.

Nair, M. N. B., Shivanna, K. R., and H. Y. M. Ram. 1995. Ethephon enhances karaya gum yield and wound healing response: a preliminary report. *Curr. Sci.* 69:809–10.

Nussinovitch, A. 1997. *Hydrocolloid applications. Gum technology in the food and other industries*. London: Blackie Academic & Professional.

Nussinovitch, A. 2003. *Water soluble polymer applications in foods*. Oxford: Blackwell Publishing.

Nussinovitch, A. 2010. *Polymer macro- and micro-gel beads: Fundamentals and applications*. New York: Springer.

Nussinovitch, A. 2010. *Plant gum exudates of the world sources, distribution, properties, and applications*. Boca Raton: CRC Press, Taylor and Francis Group.

Nussinovitch, A., and M. Hirashima. 2014. *Cooking innovations, using hydrocolloids for thickening, gelling and emulsification*. Boca Raton, FL: CRC Press, Taylor & Francis Group.

Rao, P. S., and R. K. Sharma. 1957. Indian plant gums: Composition and graded hydrolysis of karaya gum (*Sterculia urens*). *Proc. Indian Acad. Sci.* 45A:24–9.

Raymond, W. R., and C. W. Nagel. 1973. Microbial degradation of gum karaya. *Carbohydr. Res.* 30:293–312.

Shah, J. J., and R. C. Setia. 1976. Histological and histochemical changes during the development of gums canals in *Sterculia urens*. *Phytomorphology* 26:151–158.

USDA, ARS, National Genetic Resources Program. 2008. *Germplasm resources information network (GRIN)*. [Online Database] National Germplasm Resources Laboratory, Beltsville, MD. Available at: http://www.ars-grin.gov/cgi-bin/npgs/acc/display.pl.

Verma, V. P. S., and G. N. Kharakwal. 1977. Experimental tapping of *Sterculia villosa* Roxb. for gum karaya. *Indian Forester* 103:269–72.

Weiping, W., and A. Branwell. 2000. Tragacanth and karaya. In *Handbook of hydrocolloids*, ed. G. O. Philips and P. A. Williams, 231–46. Cambridge: Woodhead Publishing Ltd.

CHAPTER 9

Gum Tragacanth

9.1 INTRODUCTION

Gum tragacanth has an ancient history. It was described by Theophrastus, who was Aristotle's successor in the Peripatetic School of Philosophy, more than a few centuries before Christ. Theophrastus' interests were wide-ranging, extending from biology and physics to ethics and metaphysics. His two remaining botanical works, *Enquiry into Plants (Historia Plantarum)* and *On the Causes of Plants*, had a significant impact on Renaissance science. The term "tragacanth" came from the Greek words *tragos* (goat) and *akantha* (horn), and it most certainly refers to the curved shape of the ribbons—the best grade of commercial gum.

9.2 DISTRIBUTION AND ECONOMIC IMPORTANCE

Gum tragacanth is an exudate of *Astragalus* plants of the Leguminosae family. Its distributional range covers Iraq, Lebanon, Syria, and Turkey.

The gum exudes spontaneously from wounds or breaks in the bark of the shrubs. Gum tragacanth can be used as a food additive for emulsification and/or as a thickening agent.

9.3 WATER SOLUBILITY

The soluble ingredient, tragacanthin, dissolves in water to give a colloidal hydrosol solution; at the same time, the insoluble ingredient, bassorin, swells to a gel-like state. With a minute quantity of water, a soft, adhesive paste is formed. With more water, the paste produces a homogeneous mixture; however, after 1 or 2 days, a large proportion separates out, leaving a smaller dissolved portion. The gum is entirely insoluble in alcohol. The viscosity of the mucilage can be decreased by adding acid or alkali, particularly while heating. A number of methods can be used to eliminate lump formation and obtain a homogeneous solution, for example, strong agitation with a high-speed mixer, adding the gum gradually or, if possible, pre-wetting the gum with a wetting agent, such as glycerin, propylene glycol, or alcohol. Improved solubility has also been achieved by lyophilizing the gum. The freeze-dried gum swells rapidly and has higher initial viscosities as compared with untreated gum.

9.4 CHEMICAL CHARACTERISTICS

This is a complex gum with a molecular weight of ~800,000. Gum tragacanth is composed of a soluble arabinogalactan portion and an insoluble, but water-swellable, bassorin portion. The relative amounts of these constituents differ from 9:1 to 1:1. Hydrolysis yields arabinogalactan, xylose, fucose, galactose, rhamnose, and galacturonic acid, with traces of starch and cellulosic material. Since more than 20 dissimilar species can be utilized for gum manufacture, there are extensive differences in their composition and performance. The species which gives more viscous gum have higher proportions of fucose, xylose, galacturonic acid, and methoxyl groups in the gum and lower proportions of arabinose and nitrogenous fractions. Low-viscosity products contain extra arabinose and galactose, but lower proportions of methoxyl and galacturonic acid. Differences in the tragacanthin-to-bassorin ratio lead to dissimilar viscosities in the commercial gum. Carboxyl groups on the galacturonic acid residues are present in the form of calcium, magnesium, and potassium salts. Analysis of a representative sample of gum tragacanth yielded 70% tragacanthin, 10% soluble gum, 10% moisture, 4% cellulose, 3% starch, and 3% ash. In line with older records, the gum consists of 20% moisture, 60% tragacanthin, 8%–10% soluble gum, 3% cellulose, 2%–8% starch, 3% mineral matter, and traces of nitrogenous matter. Supplementary records have detailed 18.9% moisture, 35.9% soluble gum, 2.7% ash, and 42.4% insoluble gum; no starch was found, even though its absence is very uncommon. The insoluble fraction of gum tragacanth is dissolved by strong alkalis to form a yellow substance. The soluble portion presents the following differences from arabin found in *Acacia* gums: it does not show an acidic reaction; it does not precipitate in a solution of borax or ferric chloride; it is precipitated by both neutral and basic lead acetate, whereas acacia precipitates only in basic lead salt. Tragacanth is acidic in nature and 1 g of the gum requires 0.9 mL alkali (10 N) for neutralization. Tragacanth contains hydroxyproline in its peptides, which is most probably engaged in stabilizing the arabinogalactan structure. Studies on the structural features of tragacanthic acid can be located elsewhere.

9.5 PHYSICAL PROPERTIES

The viscosity of tragacanth solution can be reduced by heating as well as by adding acid, alkali, or sodium chloride. Viscosity is maximal at pH 8.0, and it decreases sharply at pH <4 or >6. The maximum stable viscosity has been observed at around pH 5.0. Compared to other gums, tragacanth is practically stable over a wider pH range; it is stable even at exceptionally acidic conditions at about pH 2.0. The stability may be related to the backbone resistance of the gum and to the protection afforded by the arabinofuranose side chains. Tragacanth mucilage will demonstrate amplified viscosity, if boiled or with age, but it has reduced viscosity when neutralized. Homogenizing tragacanth mucilage causes the viscosity to increase to a maximum. Viscosities of 3,500–4,600 mPa s have been observed for 1% pseudoplastic solutions. In a cold preparation, the maximum viscosity is regularly achieved after 24 hours, but it can be attained in about 2 hours by raising the temperature of the solution to around 50°C. Gum tragacanth colloidal dispersions do not demonstrate thixotropy. Tragacanth, alone, is of little value as an emulsifying agent, but a low amount of the gum in water decreases the latter's surface tension, thus enabling emulsification. Emulsions with tragacanth frequently include acacia gums. While gum arabic prevents coalescence by creating a film around the oil globules, tragacanth delays coalescence of the globules by raising the viscosity of the external phase and consequently slowing down movement of the dispersed oil phase. The elongated molecular structure of tragacanth accounts for its high viscosity. Solutions are acidic in the pH range of 5.0–6.0. To reduce the number of resistant spores from soil and airborne contaminants, an ethylene oxide gas treatment is employed on gum that is assigned for pharmaceutical products, whereas the less efficient propylene oxide is permitted for gums that are designated for food utilization.

9.6 FOOD APPLICATIONS

Tragacanth is employed in numerous low-pH products, such as salad dressings, condiments, and relishes; it serves as a stabilizer and presents a smooth mouthfeel due to its surface-active properties. Tragacanth has a wide range of properties that make it suitable for use in condiments, dressings, and sauces. The usage levels are 0.4%–0.8% of the weight of the aqueous phase and it depends upon the oil content, the use of supplementary thickeners, and the required consistency. In confections and icings, gum tragacanth serves as a water-binding agent because it has a high proportion of water-swellable (insoluble) fraction. Gum tragacanth is also used as a modifier to enhance the sensory properties of reduced-fat dairy cream, which is made from dairy cream (30 wt% fat), milk (1.5 wt% fat), and gum tragacanth. Addition of gum tragacanth to low-fat dairy cream allows fat reduction from 30 wt% to 14 wt%, with no significant changes in the sensory properties, shelf life, or packaging requirements. The fat content, the amount of gum tragacanth used, and the ripening time had major effects on the cheese-making yield, chemical characteristics, rheological characteristics, and microstructure of Iranian white cheese. As the fat content in the cheese decreased, the hardness increased and the microstructure became more compact. Adding gum tragacanth to the low-fat cheeses increased their moisture content and improved their sensory properties, but only to 0.75 g gum/kg of milk. Gum tragacanth can also reduce problems in other dairy products. For instance, serum separation in Doogh, a popular acidic dairy drink in the Middle East, happens due to low pH and aggregation of caseins. Addition of soluble tragacanthin and gum tragacanth was found to prevent serum separation at concentrations of 0.1% and 0.2%, respectively. It is possible that tragacanthin sorbs to casein and induces stabilization via electrostatic and steric repulsion. Moreover, the insoluble bassorin may assist in stabilization by increasing the viscosity.

Sweets are produced with blends of tragacanth and gum arabic to get a chewy consistency, and blends of tragacanth and gelatin are used for a chewy and cohesive texture. Tragacanth can serve as a binder in highly sweetened icings that enclose fats, to offer some pliability and to decrease moisture loss by evaporation. Flavored oil emulsions are stabilized with 0.8%–1.2% tragacanth and its blends; their shelf life is extended in the process of achieving the required combination of thickening, emulsifying, and mouthfeel properties. In frozen desserts, gum tragacanth (0.2%–0.5%) is used to control ice-crystal growth, to decrease moisture migration and ice-crystal development during storage, and to avoid color and flavor migration throughout storage and consumption. In the fillings of baked goods, the acid stability of tragacanth is exploited to obtain a creamy texture, fine clarity, and sheen. In a few applications, such as ready-to-spread icings, gum tragacanth cannot be replaced by any other gums or gum combinations.

9.7 RECIPES

9.7.1 Blue Cheese Dressing (Figure 9.1)

(Makes 600 g)
190 g (⅘ cup) mayonnaise
200 g sour cream
150 g blue cheese
15 mL (1 tbsp) vinegar*
6 g (2⅓ tsp) gum tragacanth powder
45 mL (3 tbsp) water
1 g (⅙ tsp) salt
1 g (½ tsp) black pepper
2 g (½ tsp) garlic powder

1. Crumble blue cheese with a spoon or a fork (Figure 9.1A).

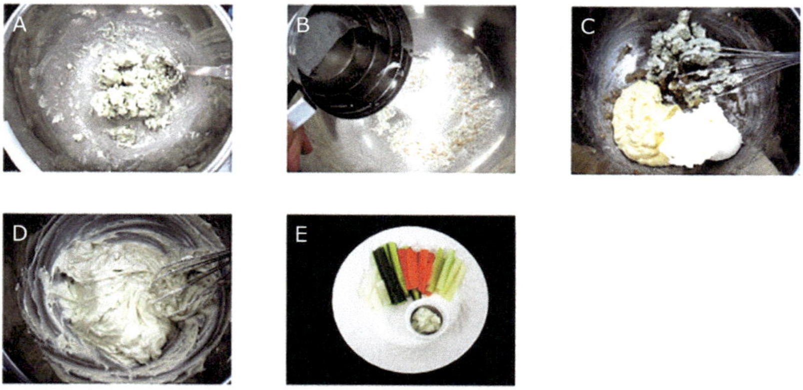

Figure 9.1 Blue cheese dressing. (A) Crumble blue cheese. (B) Add water and vinegar to the mixture of powdered ingredients. (C) Add mayonnaise, sour cream and the crumbled blue cheese. (D) Mix. (E) Serving suggestion: vegetables with blue cheese dressing.

* You can use any type of vinegar you like.

2. Mix gum tragacanth, salt, black pepper, and garlic powder in a bowl and then add water and vinegar to it (Figure 9.1B) while stirring with a hand mixer; add mayonnaise, sour cream, and the crumbled blue cheese (1) (Figure 9.1C) and mix (Figure 9.1D) (Hint 9.7.1.1).

pH=4.54

Preparation hints:

9.7.1.1 Stir vigorously with a whisk to eliminate lumps of blue cheese; the texture of the dressing should be creamy and smooth.

This recipe is good not only for salads, but also for vegetable sticks, such as carrots, cucumbers, celeries, or radishes (Figure 9.1E).

9.7.2 Sweet and Sour Sauce (Figure 9.2)

(Makes 500 g)

180 g (⅘ cup) vegetable oil
90 g (a little under ½ cup) brown sugar*
120 g (⅖ cup) ketchup
50 mL (⅕ cup) vinegar†
18 g (1 tbsp) Worcestershire sauce
6 g (2⅓ tsp) gum tragacanth powder
45 mL (3 tbsp) water
3 g (½ tsp) salt
20 g (2 tbsp) onion powder

1. Mix gum tragacanth, brown sugar, salt, and onion powder in a bowl and then add water and vinegar to it while stirring with a hand mixer (Figure 9.2A); add ketchup and Worcestershire sauce (Figure 9.2B).
2. Add vegetable oil gradually with stirring (Figure 9.2C).

---

* You can use white sugar instead.

† You can use any type of vinegar you like.

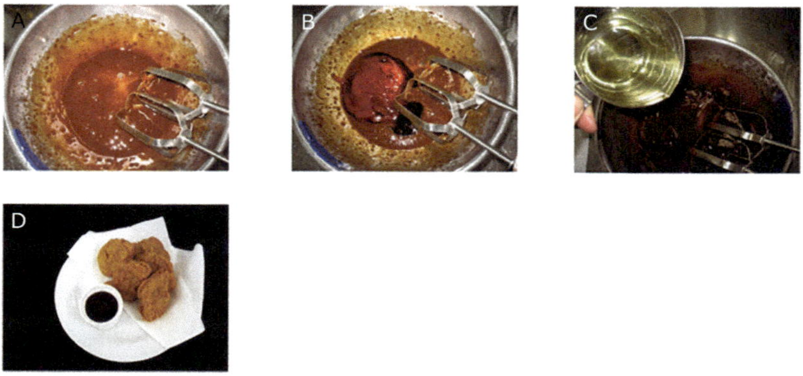

Figure 9.2 Sweet and sour sauce. (A) Mix gum tragacanth, brown sugar, salt, onion powder, water and vinegar while stirring. (B) Add ketchup and Worcestershire sauce. (C) Gradually add vegetable oil while stirring. (D) Serving suggestion: chicken nuggets with sweet and sour sauce.

pH=3.71

This recipe is good for fried dishes such as chicken nuggets (Figure 9.2D), fried fish, and French fries.

9.8 TIPS FOR THE AMATEUR COOK AND PROFESSIONAL CHEF

- Gum tragacanth and gum karaya absorb water quickly to produce viscous colloidal sols.
- Gum tragacanth can be used with other gums as a stabilizer in ice cream, ice pops, and sherbets.
- Gum tragacanth, combined with gum arabic, gives a better flavored emulsion for baking purposes.

References and Further Reading

American Pharmaceutical Association. 1946. *The committee on national formulary*, 8th edn., 67–8. Washington DC: American Pharmaceutical Association.

Anderson, D. M. W. 1989. Evidence for the safety of gum tragacanth (Asiatic *Astragalus* spp.) and modern criteria for the evaluation of food additives. *Food. Addit. Contam.* 6:1–12.

Anderson, D. M. W., and M. M. E. Bridgeman. 1985. The composition of the proteinaceous polysaccharides exuded by *Astragalus microcephalus*, *A. gummifer* and *A. kurdicus*—the sources of Turkish gum tragacanth. *Phytochemistry* 24:2301–4.

Aspinall, G. O., and J. Baillie. 1963. Gum tragacanth. Part I. Fractionation of the gum and the structure of tragacanthic acid. *J. Chem. Soc.* 1702–14.

Aspinall, G. O., and J. Baillie. 1963. Gum tragacanth. Part II. The arabinogalactan. *J. Chem. Soc.* 1714–20.

Aspinall, G. O., and V. Puvanesarajah. 1984. The hex-5-enose degradation: cleavage of 6-deoxy-6-iodo-α-D-galactopyranoseidic linkages in modified carboxyl-reduced methylated tragacanthic acid. *Can. J. Chem.* 62:2736–9.

Azarikia, F., and S. Abbasi. 2010. On the stabilization mechanism of Doogh (Iranian yoghurt drink) by gum tragacanth. *Food Hydrocolloids* 24:358–63.

Balls, E. K. 1962. *Early uses of California plants.* Berkeley: The University of California Press.

Ben-Zion, O., and A. Nussinovitch. 1997. Physical properties of hydrocolloid wet glues. *Food Hydrocolloids* 11:429–42.

Edwards, H. G. M., Falk, M. J., Sibley, M. G., Alvarez, J. B., and F. Rull. 1998. FT-Raman spectroscopy of gums of technological significance. *Spectrochimica Acta, Part A—Mol. Biomol. Spectroscopy* 54A:903–20.

Felter, H. W., and J. U. Lloyd. 1898. *King's American dispensatory*, 18th edn., 3rd revision, reprinted 1983. Portland, OR: Eclectic Medical Publications.

Ferri, C. M. 1959. Gum tragacanth. In *Industrial gums*, ed. R. L. Whistler, 511–5. New York:Academic Press.

Fowells, H. A. 1965. U.S. Department of Agriculture, Forest Service, Silvis of forest trees of the United States. Washington DC: Agriculture Handbook 271.

Glicksman, M. 1963. Natural gums. In *Kirk-Othmer encyclopedia of chemical technology*, 2nd edn., vol. 10, 741–54. New York: Wiley.

Glicksman, M. 1983. Gum tragacanth: Regulatory status, structure, properties, food applications (*Astragalus gummifer*). *Food Hydrocolloids* 2:49–60.

Howes, F. N. 1949. *Vegetable gums and resins.* Waltham, MA: Chronica Botanica Company.

Imeson, A. P. 1992. Natural plant exudates. In *Thickening and gelling agents for food*, ed. A. P. Imeson. London: Blackie Academic & Professional.

Levi, R. S. 1955. A study of lyophilized tragacanth, Ph. D. thesis, University of Florida, Gainesville.

Levy, G., and T. W. Schwarz. 1958. Tragacanth solutions. I. The relation of method of preparation to the viscosity and stability. *J. Am. Pharmaceut. Assoc.—Scientific Edn.* 47: 451–4.

Levy, G., and T. W. Schwarz. 1958. Tragacanth solutions. II. The determination of thickening capacity and stability. *Drug Standards* 26:153.

Lewis, W. H., and M. P. F. Elvin-Lewis. 1977. *Medical botany*. New York: John Wiley & Sons.

Mantell, C. L. 1947. *The water-soluble gums*. New York: Reinhold Publ. Corp.

Nasirian, R., Vaziri, A., Safekordi, A. A., and M. Ardjmand. 2010. Use of gum tragacanth in reduction of dairy cream fat. *Milchwissenschft-Milk Sci. Int.* 65:49–52.

Nussinovitch, A. 1997. *Hydrocolloid applications. Gum technology in the food and other industries*. London: Blackie Academic & Professional.

Nussinovitch, A. 2003. *Water soluble polymer applications in foods*. Oxford, UK: Blackwell Publishing.

Nussinovitch, A. 2010. *Polymer macro- and micro-gel beads: Fundamentals and applications*. New York: Springer.

Nussinovitch, A. 2010. *Plant gum exudates of the world sources, distribution, properties, and applications*. Boca Raton, FL: CRC Press, Taylor and Francis Group.

Nussinovitch, A., and M. Hirashima. 2014. *Cooking innovations, using hydrocolloids for thickening, gelling and emulsification*. Boca Raton, FL: CRC Press, Taylor & Francis Group.

Rahimi, J., Khosrowshahi, A., Madadlou, A., and S. Aziznia. 2007. Texture of low-fat Iranian white cheese as influenced by gum tragacanth as a fat replacer. *J. Dairy Sci.* 90:4058–70.

Schaub, K. 1958. Rheological standardization of tragacanth and the evaluation of the emulsifying powers of acacia. *Pharmaceutica Acta Helvetica* 33:797–851.

Stauffer, K. R. 1980. Gum tragacanth. In *Handbook of water-soluble gums and resins*, ed. R. L. Davidson, 111–31. New York: McGraw-Hill.

Sybil, P. J., and F. Smith. 1945. The chemistry of gum tragacanth. Part I. Tragacanthic acid. *J. Chem. Soc.* 739–46.

Sybil, P. J., and F. Smith. 1945. The chemistry of gum tragacanth. Part II. Derivatives of D- and L-fucose. *J. Chem. Soc.* 746–8.

Sybil, P. J., and F. Smith. 1945. The chemistry of gum tragacanth. Part III. *J. Chem. Soc.* 1945:749–51.

Weiping, W., and A. Branwell. 2000. Tragacanth and karaya. In *Handbook of hydrocolloids*, ed. G. O. Philips and P. A. Williams, 231–46. Cambridge: Woodhead Publishing Ltd.

CHAPTER 10

Inulin

10.1 INTRODUCTION

A carbohydrate that was isolated in 1804 from *Inula helenium* was given the name of inulin in 1811. Inulins can be found in a variety of crops. The inulin content (as % of fresh weight) for chicory, Jerusalem artichoke, onion, garlic, barley, and wheat is 15–20, 3–10, 1–8, 9–16, 0.5–1.5, and 1–4, respectively. Inulin can also be found in crops such as yacon tubers from Japan and South America and murnong from Australia. Inulin is composed of 2–60 fructose units, with one terminal glucose molecule. The generic inulin includes all the β-(2,1) fructans. In general, inulins are polydisperse mixtures of fructan chains with varying degrees of polymerization (DPs), with the DP depending upon the crop from which they originate.

10.2 PRODUCTION

Commercial production of inulin from chicory roots involves the slicing of chicory roots, followed by hot water extraction. The diffusion fluid is transformed into a clear medium by liming and carbonation.

Another approach involves precipitation and flocculation of the proteins and cell components, in slightly acidic conditions, followed by filtration. A variety of separation techniques can be used to refine native inulin solutions. Ion exchange at low temperatures (to avoid hydrolysis at low pH) removes salts and a major proportion of the associated color. After these stages are completed, a non-viscous, slightly acidic, nearly colorless, and almost completely demineralized inulin "juice" is obtained. One more step—the use of activated carbon—is required to remove its bitter taste and residual chicory color. The inulin "juice" is concentrated by evaporation and is spray-dried to produce a dry powder. Crystallization can be employed to separate native inulin into short- and long-chain fractions which are then dried separately.

10.3 PHYSICAL AND CHEMICAL PROPERTIES

The chain-length distribution of inulin is responsible for its successful solubilization in water, which increases with decreasing chain length. Oligofructose has a DP of up to 10 and an average of ~4. Long-chain inulin has an average DP of at least 20. As a result of its low DP, oligofructose is marketable as a syrup, with a dry matter content of 75%. Such solutions are clear at temperatures as low as 5°C. The longer the storage time, the higher the likelihood of obtaining a cloudy solution; this is possibly due to the precipitation of long-chain inulins. The long-term solubility of inulin at different temperatures was assessed. For native inulin solutions stored at 5°C and 20°C, clear solutions can be obtained after 13 weeks of storage, by utilizing concentrations <2% and <2.5%, respectively. For long-chain inulin, under the same storage temperatures, concentrations <1.0% and <1.5%, respectively, were required. This information is valuable for the beverage industry and for other fluid preparations that contain inulin, since there is no interest in changing the clarity of the fluid due to precipitated inulin particles.

The viscosity of native inulin as well as of long-chain inulin is not high. Moreover, the two types of inulin do not differ too much. A 20% solution of long-chain inulin has a viscosity of ~4.5 mPa s at 20°C. Up to 140°C, inulin is stable; at higher temperatures it breaks down.

Hydrolysis will not occur at pH > 4.0. At pH < 4.0, hydrolysis occurs and is dependent on pH, the dry matter content of the product, and the time-temperature combinations.

Inulin's ability to bind water is lesser than that of other hydrocolloids. At a concentration >15%, inulin can produce a cream or gel. The gel can be regarded as a particle gel, somewhat similar to that formed by several starches. At concentrations <15%, solutions with lower viscosities are obtained. Inulin has a high number of hydroxyls; as such, it competes for water and, therefore, it might influence the solubility of other water-binding ingredients. Addition of inulin reduces the viscosity of starch solutions more than the same quantity of added sugar. At high additions of inulin, viscosity is much lower. Addition of inulin to κ-carrageenan gel increases its brittleness. A typical procedure for producing an inulin gel includes its dissolution in tap water, heating at 72°C for 15 minutes while stirring, and then storing the solution at 5°C for at least 6 hours. The water is retained in the gel network that is formed. The rheological properties of the gel resemble that of fat and, therefore, inulin is of interest for products that have low or zero fat, but require the structure of a fattier substance. The gel features depend upon chain-length distribution, concentration, preparation temperature, amount of shear, and the option to add inulin seed crystals. Higher gel strength can be achieved by using long-chain inulins. In addition, at least 20% native inulin is required to form a gel. It has been shown that inulins with DP > 10 are part of the gel structure while inulins with smaller DPs remain dissolved. Maximal firmness can be achieved by heating the inulin solution at high temperatures to obtain complete hydration, by applying shear after pasteurization or during cooling, and by including seed crystals.

10.4 NUTRITIONAL AND HEALTH ASPECTS

Inulins reach the small intestine unharmed due to their glycosidic bonds' resistance to gastric acids and the lack of digestive enzymes that can split the molecule. At the small intestine, inulins are fermented and short-chain fatty acids—acetic, propionic, and butyric—are produced and absorbed. Inulins are low in calories (~1.5 kcal/g). They have a prebiotic effect, i.e., their fermentation in the large intestine leads to a change in microflora that leads to enhanced growth of healthy bacteria. Fermentation of inulins can have an influence on both fiber functions and health-related issues. The fiber effects lead to increased defecation and stool bulking. The non-digestibility of the fiber contributes to a low glycemic effect. In other words, inulin can be used as a suitable ingredient for the development of food with a low glycemic response and as a food for diabetics. As a result of inulin consumption and, perhaps, due to the localized lowering of pH by fermentation, the absorption of calcium and magnesium increases, which can influence bone density. Another beneficial effect is a reduction in the level of serum lipids, leading to a possible decrease in the risk of cardiovascular diseases. Inulin might have a positive effect on satiety and energy intake. An influence on immune health might also be expected due to the increase in bifidobacteria and lactobacilli. Moreover, inulins might lower the risk of colon cancer. However, many more human trials are required to confirm these potential benefits.

10.5 FOOD APPLICATIONS AND REGULATORY STATUS

Inulin gels can be used to replace fats. Inulins can be used in dairy, baked, and other miscellaneous products, such as spreads, cereals, dressings, chocolate, meat products, and pasta fillings. In beverages, the addition of inulin to traditional thickeners, such as guar gum, pectin, and xanthan gum increases the product homogeneity. In dairy products, the inclusion of long-chain inulins (DP ~20–25) improves the mouthfeel and the texture and reduces syneresis. Creamier dairy products are obtained due to the interactions of inulins with whey proteins and caseinates. The addition of native inulin (DP 9–11) improves the mouthfeel and might improve foam stabilization.

In baked goods, long-chain inulins can be used to replace flour and enhance crispiness. The native inulins might also replace flour, improve texture and foam stabilization, and enhance crispiness in baked items.

The possibility of replacing 25%–100% of the fat, in baked yeast recipes, with inulin was evaluated, and the organoleptic properties of the newly created products were assessed. Inulin did not bring about any negative organoleptic changes in the examined confectionery products. Replacing half of the fat with inulin had a positive effect on the scent and the consistency of the modified products. A chocolate cake formulation was attempted, in which wheat flour was partially replaced with inulin and/or yacon meal. The yacon meal was prepared from the plant's tuberous roots. Formulations where 20% of the wheat flour is replaced with only yacon meal showed the best values for hardness, cohesiveness, and specific volume. A formulation where 40% of the wheat flour is replaced with yacon meal and 6% is replaced with inulin showed values similar to that of the optimal formulation, for the three mentioned responses. However, the former had a greater content of fructooligosaccharides and inulin. Consequently, both formulations may give useful functional food with comparable physical parameters. Orange cakes that were treated with inulin and oligofructose were studied to justify a prebiotic claim (minimum of 3 g of fructans in a 60 g serving of cake). The cakes with prebiotics presented greater crust brownness, dough beigeness, hardness, and stickiness than the standard cake, as well as lower crumbliness. Sensory acceptability was similar for the three cakes and was higher as compared with three commercial cakes, and the preference tests demonstrated that cakes with prebiotics were preferred to commercial ones.

Addition of 7.5% inulin to fat and dairy spreads resulted in tastier products with a better structure and spreadability, for example, in sandwiches. Marketing opportunities for lower-fat cheeses exist, but such cheeses have to be organoleptically satisfactory and they have to display textural properties that are comparable to those of the full-fat products. Inulin has been effectively added, as a partial fat substitute, to Mozzarella cheese and fermented sausages. When incorporated as a fat mimetic into natural cheese, at levels of up to 10 g/100 g (w/w), there was no effect on cheese aroma. Inulin can replace up to 63% of the fat in imitation cheese, and the preferred means of addition is in the form of a heated inulin solution. Another study found that it is possible to add up to 6% inulin in the form of a gel to bologna-type sausages with citrate in the formulation, producing a significant reduction in the energy content (22%) without negatively affecting the sensory quality. Low-fat, dry, fermented sausages were manufactured with a fat content of ~50% and 25% of the original quantity. The batch with the smallest proportion of fat was less tender, less springy, and gummier than that with the highest proportion. When the 25% batch was supplemented with different amounts of inulin, on the whole, improved sensory properties were achieved owing to a softer texture, tenderness, springiness, and adhesiveness, which more closely resembled that of the conventional high-fat sausage.

Inclusion of 5% inulin in a mayonnaise dressing increases its stability and improves its texture. Long-chain inulin increases the perceived creaminess of skimmed yogurt. Inulin might also contribute to the improved sensory qualities of low-fat and low-sugar ice creams. Alongside its health benefits, inulin is also considered to have functional properties, such as the ability to replace fat or sugar without adversely affecting flavor. The fat-substituting property of inulin is based on its ability to stabilize the structure of the aqueous phase, which creates an improved "creamy" mouthfeel. The inclusion of inulin has important effects on the structure and texture of ice cream–yogurt mixes. Such differences may be caused by the water-binding capacity of inulin and by the way in which it gets incorporated into the foam mixture during the freezing process. More practically, the addition of 5% inulin appears to produce acceptable products that have the required attributes to influence consumer acceptability of reduced-fat frozen desserts.

A prebiotic is defined as "a non-digestible food ingredient that beneficially affects the host, by selectively stimulating the growth and/or activity of one or a limited number of beneficial bacteria present in the colon that can improve the host's health." Prebiotics have the potential to boost not only the viable counts of probiotics, but also their metabolic activity by supplying fermentable substrates. Inulin-type fructans are a group of prebiotics found in plants. Inulins extracted from Jerusalem artichoke tubers were tested for their prebiotic and fat-replacement properties and were compared to commercial inulin, Beneo ST, which was extracted from chicory roots. Yogurt made with Jerusalem artichoke inulin showed better retention of probiotic viability (*Bifidobacterium bifidum* and *Lactobacillus acidophilus*) than that made with commercial inulin. Textural and rheological studies have indicated that Jerusalem artichoke inulin can function as a fat replacer in low-fat yogurt while mimicking the properties of full-fat yogurt. Another study described the preparation of a frozen yogurt containing low fat and no added sugar. This product was analyzed for its physical and chemical properties. With the addition of inulin and isomalt, viscosity increased by 19%–52% as compared to that of the reduced-fat control. The sensory properties of the sample, containing 6.5% inulin and 6.5% isomalt, were similar to those of the controls. Inulin gel can partially replace pastry margarine in low-fat cakes and can replace 25% of the fat in biscuit fillings. Due to its ability to bind water, inulin might be included in wafers to create a product that is crisp even at higher moisture content. Inulin is a part of our daily diet and, therefore, regulations regarding its inclusion differ from country to country. It has no E-number, but can be used in food categories and has been recognized as a dietary fiber. Its fiber claims can be located in EU 1924/2006. In general, each claim, be it structural, health or otherwise, is detailed in each country or continent's regulations.

10.6 RECIPES

10.6.1 Ketchup (Figure 10.1)

(Makes 180 g)

130 g (½ cup) tomato purée

25 g (2 tbsp) brown sugar*

19 mL (1⅓ tbsp) vinegar†

15 g (¾ tbsp) inulin powder (DP < 10)

7 mL (1⅖ tsp) water

4 g (⅔ tsp) salt

Black pepper, onion powder, garlic powder, celery seed, red pepper powder (optional)

* White sugar can be used instead.

† Any type of vinegar can be used.

FIGURE 10.1 KETCHUP. (A) MIX BROWN SUGAR, INULIN, AND SALT. (B) MIX TOMATO PURÉE, VINEGAR, AND WATER INTO THE POWDER MIXTURE. (C) STIR UNTIL HOMOGENIZED.

1. Mix brown sugar, inulin, and salt in a bowl (Figure 10.1A).
2. Add tomato purée, vinegar, and water to the powder mixture (1) and mix with a hand mixer for 1 minute (Figure 10.1B).
3. Pour mixture (2) into a pan, add black pepper, onion powder, garlic powder, celery seed or red pepper powder to taste, and heat while stirring with a whisk or a hand mixer at 70°C to bring homogeneity (Figure 10.1C) (Hint 10.6.1.1). Heat to 90°C.
4. Pour the ketchup (3) into a glass bottle or container and cool (Hint 10.6.1.2).

pH = 3.87

Preparation hints:

10.6.1.1 Vigorous stirring is recommended.

10.6.1.2 Store in the refrigerator.

This recipe is easily made using tomato purée and has excellent textural properties. You can reduce the sugar content slightly; you can adjust the sour and sweet taste by altering the vinegar and sugar contents, respectively.

10.6.2 LOW-FAT CHOCOLATE MOUSSE (FIGURE 10.2)

(Serves 5)

20 g (1 tbsp) inulin powder (DP < 8)

60 mL (4 tbsp) water

100 g bitter chocolate (60%–80% cacao) (Hint 10.6.2.1)

2 egg yolks
2 egg whites
27 g (3 tbsp) granulated sugar
100 g (⅖ cup) heavy cream
Vanilla flavoring

1. Mix inulin and water in a cup (Figure 10.2A) and refrigerate overnight (Figure 10.2B) (Hint 10.6.2.2).
2. Chop chocolate, place it in a bowl, put the bowl in a hot-water bath (80°C–90°C), and stir occasionally until the chocolate melts (45°C–50°C) (Figure 10.2C).
3. Mix egg yolk into the melted chocolate (2) (Figure 10.2D).

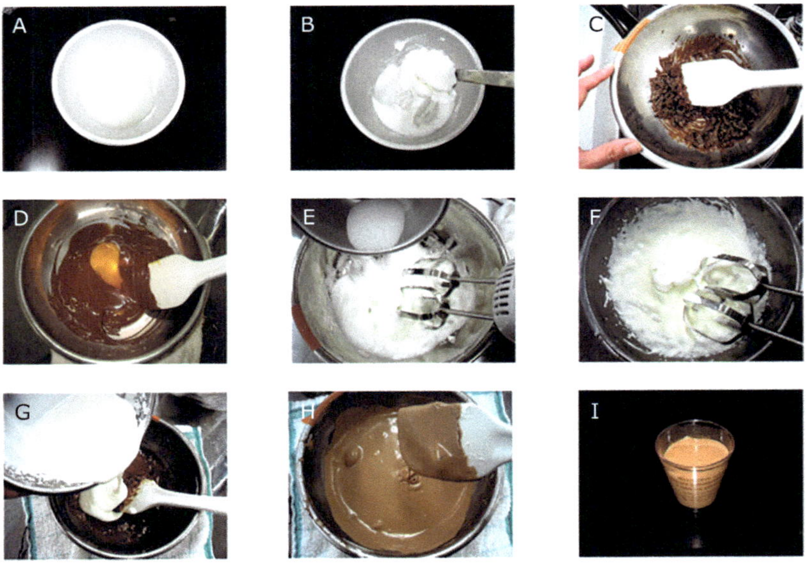

Figure 10.2 Low-fat chocolate mousse. (A) Mix inulin and water. (B) Refrigerate overnight. (C) Chop chocolate and melt it. (D) Add egg yolk to the melted chocolate and mix. (E) Whisk egg white until stiff and gradually whisk in granulated sugar. (F) Whip heavy cream and then mix in the inulin dispersion. (G) Add the whipped egg whites, heavy cream, and vanilla flavoring to the chocolate mixture. (H) Mix the ingredients well. (I) Pour into molds and chill.

4. Whisk egg white with a hand mixer until stiff peaks form and gradually add granulated sugar (Figure 10.2E).
5. Whisk heavy cream with a hand mixer until stiff peaks form and add the inulin dispersion (1) to it (Figure 10.2F) while mixing well.
6. Add the whipped egg whites (4) to the chocolate mixture (3) in a few steps, gently folding them in; add the whipped cream (5) and vanilla flavoring (Figure 10.2G) and mix well (Figure 10.2H).
7. Pour the chocolate mousse (6) into molds and cool (Figure 10.2I).

pH = 6.32

Preparation hints:

10.6.2.1 Add 5 g instant coffee for a more bitter taste.

10.6.2.2 When cooled, the inulin forms a microcrystalline structure. As a result, the inulin dispersion has a cream-like texture.

This recipe has a smooth and airy texture, and the chocolate sets at a lower temperature without the need of a gelling agent. The inulin dispersion substitutes for 50% of the heavy cream in this recipe. You can use inulin to replace all of the heavy cream, but the inulin dispersion cannot be whipped, and the smooth and airy texture will be weakened.

10.6.3 *PA-JUN* (KOREAN PANCAKE WITH GREEN ONION) (FIGURE 10.3)

(Serves 5)

100 g (a little under ¾ cup) flour

2 g (⅓ tsp) inulin powder

1 egg

200 g (⅘ cup) soup stock made of small, dried fish

100 g *kimchi*

100 g onion

50 g carrot

100 g green onion

5 slices of bacon

Sesame oil

A
- 30 mL (1 ⅖ tbsp) soy sauce
- 30 mL (2 tbsp) vinegar
- 5 g (1⅔ tsp) white sugar
- 5 g (1⅔ tsp) sesame
- 25 g green onion

1. Cut *kimchi*, green onion, and bacon into bite-sized pieces. Then slice onion and carrot thinly (Figure 10.3A).
2. Mix flour, inulin, egg, and soup stock (Figure 10.3B) and add the cut ingredients (1) (Figure 10.3C).
3. Heat a frying-pan, add sesame oil, and pour batter (2) into the frying pan; fry it, turn over to fry the other side, while adding sesame oil occasionally (Hint 10.6.3.1).
4. Make a dipping sauce using ingredients in A. Slice green onion diagonally and mix soy sauce, vinegar, white sugar, sesame, and the sliced green onions in a bowl.
5. Cut the pancake (*pa-jun*) (3) into bite-sized pieces and serve alongside the dipping sauce (4) (Figure 10.3D).

pH = 6.28

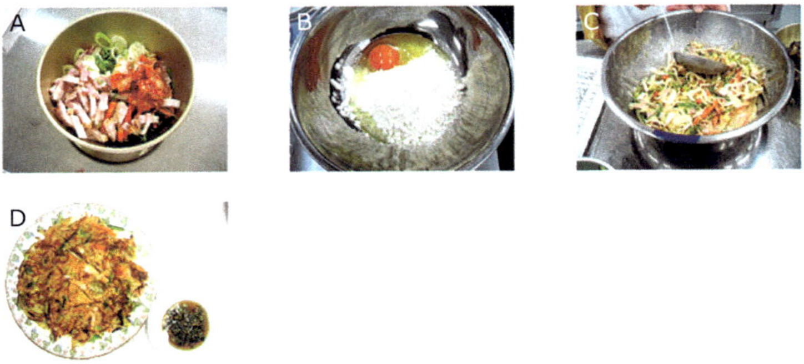

FIGURE 10.3 *PA-JUN* (KOREAN PANCAKE WITH GREEN ONION). (A) CUT *KIMCHI*, GREEN ONION, AND BACON, THEN SLICE ONION AND CARROT. (B) MIX FLOUR, INULIN, EGG, AND SOUP STOCK. (C) ADD THE CUT INGREDIENTS. (D) CUT THE PANCAKE AND SERVE ALONGSIDE DIPPING SAUCE.

Preparation hints:

10.6.3.1 Adding sesame oil produces a good smell and a crispy texture.

This recipe has broad applications, as many different kinds of vegetables, seafood, and meat can be used in it. It is usually served as a snack, appetizer, or side dish.

10.7 TIPS FOR THE AMATEUR COOK AND PROFESSIONAL CHEF

- Inulin will break down at temperatures >140°C, but in powdered form it is stable at higher temperatures.
- At pH < 4.0, hydrolysis may occur.
- Native and long-chain inulins can be used at high dosages of up to 20% in dairy mousse, and miscellaneous chocolates and dressings.
- Inulin will reduce the smell of vegetables in vegetable juice and the fishy smell of boiled fish.
- Inulin will reduce the sour taste in marinated food.
- Adding inulin to fritter or *tempura* batter gives a crispy texture.
- Adding inulin to pancake batter or bread dough produces a fluffy texture.

REFERENCES AND FURTHER READING

Archer, B. J., Johanson, S. K., Devereux, H. M., and A. L. Baxter. 2004. Effect of fat replacement by inulin or lupin-kernel fibre on sausage patty acceptability, post-meal perceptions of satiety and food intake in men. *Br. J. Nutr.* 91:591–9.

Beylot, M. 2005. Effects of inulin-type fructans on lipid metabolism in man and in animal models. *Br. J. Nutr.* 93:S163–8.

Bot, A., Erle, U., Vreeker, R., and W. G. M. Agterof. 2004. Influence of crystallisation conditions on the large deformation rheology of inulin gels. *Food Hydrcolloids* 18:547–56.

Causey, J. L., Feirtag, J. M., Gallaher, D. D., Tungland, B. C., and J. L. Salvin. 2000. Effects of dietary inulin on serum lipids, blood glucose, and the gastrointestinal environment in hypercholesterolemic men. *Nutr. Res.* 20:191–201.

Den Hond, E., Geypens, B., and Y. Ghoos. 2000. Effect of high performance chicory inulin on constipation. *Nutr. Res.* 20:731–6.

El-Nagar, G., Clowes, G., Tudorica, C. M., Kuri, V., and C. S. Brennan. 2002. Rheological quality and stability of yog-ice cream with added inulin. *Int. J. Dairy Technol.* 55:89–93.

European Commission. 2006. Regulation for nutrition and health claims made on foods (EU 1924/2006), available at: http://eur-europa.eu/LexUriServ/site/en/oj/2007 l_012/1_01220070118en000300018.pdf.

Franck, A. 2000. *Prebiotics and probioteics ingredients handbook.* Reading: Leatherhead Food Publishing.

Fuji Nihon Seito Co. Inulin, *fructfiber*, available at http://www.fnsugar.co.jp/guide/inulin/index.html, in Japanese.

Hennelly, P. J., Dunne, P. G., Sullivan, M. O., and E. D. O'Riordan. 2006. Textural, rheological and microstructural properties of imitation cheese containing inulin. *J. Food Eng.* 75:388–95.

Isik, U., Boyacioglu, D., Capanoglu, E., and D. Nilufer Erdil. 2011. Frozen yogurt with added inulin and isomalt. *J. Dairy Sci.* 94:1647–56.

Kim, S.-H., Lee, D. H., and D. Meyer. 2007. Supplementation of baby formula with native inulin has a prebiotic effect in formula-fed babies. *Asia Pacific J. Clin. Nutr.* 16: 172–7.

Kim, Y., Faqih, M. N., and S. S. Wang. 2001. Factors affecting gel formation of inulin. *Carbohydr. Polym.* 46:135–45.

Kip, P., Meyer, D., and R. H. Jellema. 2006. Inulins improve sesoric and textural properties of low fat yoghurts. *Int. Dairy J.* 16:1098–103.

Kusuma, G. D., Paseephol, T., and F. Sherkat. 2009. Prebiotic and rheological effects of Jerusalem artichoke inulin in low-fat yogurt. *Aust. J. Dairy Technol.* 64:159–63.

McDevitt-Pugh, M., and B. Peters. 2005. Weight management supported by tasty inulin-based products. *Agro Food Industry Hi-Tech* 16:36–7.

Mendoza, E., Garcıa, M. L., Casas, C., and M. D. Selgas. 2001. Inulin as fat substitute in low fat, dry fermented sausages. *Meat Sci.* 57:387–93.

Meyer, D., and M. Stasse-Wolthus. 2006. Inulin and bone health. *Curr. Top. Nutraceut. Res.* 4:211–26.

Michalak-Majewska, M., Zukiewicz-Koc, W., and J. Kalbarczyk. 2008. The possibility of applying inulin as fat substitute in bakery products. *Bromatologia i Chemia Toksykologiczna* 41:616–20.

Moscatto, J. A., Borsato, D., Bona, E., de Oliveira, A. S., and M. C. de Oliveira Hauly. 2006. The optimization of the formulation for a chocolate cake containing inulin and yacon meal. *Int. J. Food Sci. Technol.* 41:181–8.

Nowak, B., Vonmueffling, T., Grotheer, J., Klein, G., and B. M. Watkinson. 2007. Energy content, sensory properties and microbiological shelf life of German bologna-type sausages produced with citrate or phosphate and with inulin as fat replacer. *J. Food Sci.* 72:S629–38.

Nussinovitch, A. 1997. *Hydrocolloid applications. Gum technology in the food and other industries.* London: Blackie Academic & Professional.

Nussinovitch, A. 2003. *Water soluble polymer applications in foods.* Oxford: Blackwell Publishing.

Nussinovitch, A. 2010. *Plant gum exudates of the world sources, distribution, properties and applications.* Boca Raton, FL: CRC Press, Taylor & Francis Group.

Nussinovitch, A., and M. Hirashima. 2014. *Cooking innovations, using hydrocolloids for thickening, gelling and emulsification.* Boca Raton, FL: CRC Press, Taylor & Francis Group

Pagliarini, E., and N. Beatrice. 1994. Sensory and rheological properties of low-fat "pasta filata" cheese. *J. Dairy Res.* 61:299–304.

Pedersen, A., Sandstrom, B., and Van Amelsvoort, J. M. M. 1997. The effect of ingestion of inulin on blood lipids and gastrointestinal symptoms in healthy females. *Br. J. Nutr.* 78:215–22.

Schaller-Povolny, L. A., and D. E. Smith. 1999. Sensory attributes and storage life of reduced fat ice cream as related to inulin content. *J. Food Sci.* 64:555–9.

Sensus, D. M. 2009. Inulin. In *Handbook of hydrocolloids*, ed. G. O. Phillips and P. A. Williams, 829–45. Boca Raton: CRC Press and Oxford: Woodhead Publishing Limited.

Silva, R. F. 1996. Use of inulin as a natural texture modifier. *Cereal Foods World* 41:792–4.

Van Loo, J., Coussment, P., De Leenheer, L., Hoebregs, H., and G. Smits. 1995. On the presence of inulin and oligofructose as natural ingredients in the western diet. *Crit. Rev. Food. Sci. Nutr.* 35:535–52.

Volpini-Rapina, L. F., Sokei, F. R., and A. C. Conti-Silva. 2012. Sensory profile and preference mapping of orange cakes with addition of prebiotics inulin and oligofructose. *LWT—Food Sci. Technol.* 48:37–42.

CHAPTER 11

Larchwood Arabinogalactan

11.1 INTRODUCTION

Arabinogalactan is a water-soluble gum that is found at remarkably high concentrations (up to 35%) in the heartwood of all larch species. Arabinogalactans are also found in minor amounts in many other trees, plants and plant seeds, and in the cell walls of certain bacteria. Although a description of this hydrocolloid first appeared in texts in 1898, arabinogalactan was not produced for the commercial market until 1964. It has numerous exclusive features, such as thorough miscibility with water and low viscosity at a high concentration of dissolved solids. These distinctive characteristics make this polysaccharide beneficial for the carbon black, mining, printing, and food industries, where exciting uses for it are being created.

11.2 EXUDATE APPEARANCE AND DISTRIBUTION

Larchwood gum goes by the general names of arabinogalactan, Stractan, wood gum, and wood sugar (*Larix* of North America). Its most important uses for trade and industry are in wood manufacture and folk medicine. Its distributional range covers Canada and the United States. The gum gathers in masses beneath the bark after injury. The excrescence can be picked off manually and is generally around 95% pure. The supply of this gum is limited and, thus, gathering it is difficult and labor-intensive.

11.3 GUM WATER SOLUBILITY

Larch gum is readily soluble in water. A clear aqueous solution containing up to 75% arabinogalactan, with pH 4.0–4.5, can be prepared. Solubility increases with increasing temperature. However, these exceedingly concentrated solutions are atypical due to their Newtonian flow properties. Solutions can tolerate the addition of absolute alcohol, without precipitating out, at 70% of their own volume, if the alcohol is added gradually with constant stirring.

11.4 GUM CHEMICAL CHARACTERISTICS AND PHYSICAL PROPERTIES

This highly branched copolymer contains L-arabinose and D-galactose in a ratio of 1:6 and it is made up of two portions with average molecular weights of 16,000 and 100,000. Arabinogalactan contains 16% volatile pinene and limonene. Broad research has been done into the molecular structures of a range of arabinogalactans that are obtained from the *Larix* species. Larch gum decreases the surface tension of aqueous solutions and the interfacial tension in water–oil mixtures and is, therefore, a successful emulsifying agent. As an outcome of these properties, larch gum has been utilized in food and can serve as a replacement for gum arabic. It has also been evaluated as wet glue for skin applications.

11.5 COMMERCIAL AVAILABILITY OF THE GUM AND APPLICATIONS

An effort to make commercial use of larch gum was launched by International Chemical Products, Montana, toward the beginning of the 1920s. The gum was extracted, hydrolyzed, and oxidized to mucic acid by treatment with nitric acid. The mucic acid was employed as a baking powder constituent, but there was no reason to manufacture a gum that could compete with gum arabic. The business enterprise did not succeed, probably because the process was too expensive. In the early 1960s, a countercurrent hot-water extraction system was developed, and the gum was manufactured commercially by the St Regis Paper Co. under the trade name Stractan. The potential yearly production of this gum at the manufacturing facility was 10,000 tons, and the gum was produced from the wood remains that came from the lumber industry. The product could not compete with gum arabic, and commercial manufacture was done in small batches. This gum was used for specific purposes such as offset lithography, food, pharmaceuticals, paint, and ink. The gum is approved in the United States by the Food and Drug Administration (21 CFR 172.610) as a food additive that can act as an emulsifier, stabilizer, and binding or bodying agent. The best commercial source of the gum is the heartwood of *Larix occidentalis,* which has 8%–25% gum in the lower stem of the tree, on a dry wood basis. Other *Larix* species may also yield commercial quantities of the gum.

11.6 RECIPES

11.6.1 Sugar Snap Cookies (Figure 11.1)

(Makes 12 cookies, 9 cm in diameter)

55 g (¼ cup) butter
50 g (¼ cup) shortening
120 g (¾ cup) white sugar
2 egg yolks (60 g)

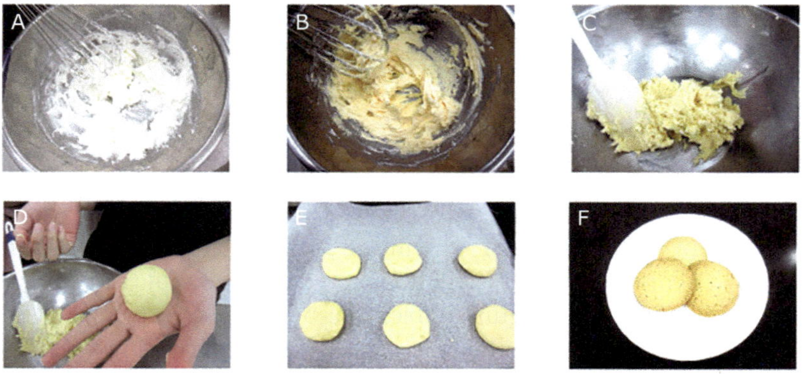

FIGURE 11.1 SUGAR SNAP COOKIES. (A) CREAM BUTTER, SHORTENING, AND SUGAR. (B) ADD EGG YOLK AND STIR. (C) ADD ALL-PURPOSE FLOUR, LARCHWOOD ARABINOGALACTAN POWDER, BAKING SODA, SALT, AND VANILLA ESSENCE AND MIX. (D) ROLL DOUGH INTO BALLS. (E) PLACE ON THE BAKING TRAY. (F) SERVE.

130 g (1 cup) all-purpose flour
2.5 g (1 tsp) larchwood arabinogalactan powder
2 g (½ tsp) baking soda
1 g (⅙ tsp) salt
Vanilla essence

1. Preheat oven to 150°C and cover baking tray with baking paper.
2. Cream butter, shortening, and sugar in a bowl using a whisk (Figure 11.1A).
3. Stir egg yolks into the creamed mixture (2) with a whisk (Figure 11.1B).
4. Add all-purpose flour, larchwood arabinogalactan powder, baking soda, salt, and vanilla essence to the mixture (3) and mix with a spatula (Figure 11.1C).
5. Roll dough (4) into 5 cm balls (30 g each) (Figure 11.1D) and place on the baking trays (5) (Figure 11.1E) (Hint 11.6.1.1).
6. Bake the balls (5) for 20 minutes and cool on wire racks (Figure 11.1F).

pH of dough = 6.2

Preparation hints:

11.6.1.1 Leave some distance between the balls because they will spread while baking (8–9 cm).

Adding larchwood arabinogalactan to cookies gives them a crispy texture that lasts for a few weeks.

11.7 TIPS FOR THE AMATEUR COOK AND PROFESSIONAL CHEF

- Larchwood arabinogalactan improves the texture of baked goods by reducing the stickiness of the dough.
- Larchwood arabinogalactan lowers water activity and aids in flavor and oil retention in confections.

References and Further Reading

Acree, S. F. 1931. Products of hydrolysis from western larch wood. U.S. Patent No. 1,816,315.

Acree, S. F. 1937. Galactan product. U.S. Patent No. 2,073,616.

Adams, M. F., and B. V. Ettling. 1973. Larch arabinogalactan. In *Industrial gums*, ed. R. L. Whistler, 415–27. New York: Academic Press.

Balls, E. K. 1962. *Early uses of California plants*. Berkeley: The University of California Press.

Beuth, J., Ko, H. L., Schirrmacher, V., Uhlenbruck, G., and G. Pulverer. 1988. Inhibition of liver metastasis in mice by blocking hepatocyte lectins with arabinogalactan. *Clin. Exp. Metastasis* 6:115–20.

Borgin, G. L. 1949. Molecular properties of water-soluble polysaccharides from Western Larch. *J. Am. Chem. Soc.* 71:2247–8.

Bouveng, H. O., and B. Lindberg. 1958. Studies on arabinogalactan. II. Fractionation of the arabogalactan from *Larix occidentalis* Nutt. A methylation study of one of the components. *Acta Chem. Scand.* 12:1977–84.

Churms, S. C., Merrifield, E. H., and A. M. Stephen. 1978. Regularity within the molecular structure of arabinogalactan from Western larch (*Larix occidentalis*). *Carbohydr. Res.* 64:C1–2.

Clarke, A. E., Anderson, R. L., and B. A. Stone. 1979. Form and function of arabinogalactans and arabinogalactan-proteins. *Photochemistry* 18:521–40.

Glicksman, M. 1963. Natural gums. In *Kirk-Othmer encyclopedia of chemical technology*, 2nd edn., vol. 10, 741–54. New York: Wiley.

Glicksman, M., and R. E. Schachat. 1966. Gelatin-type jelly dessert mix. U.S. Patent No. 3,264,114.

Haq, S., and G. A. Adams. 1961. Structure of an arabinogalactan from Tamarack (*Larix laricina*). *Can. J. Chem.* 39:1563–73.

Howes, F. N. 1949. *Vegetable gums and resins.* Waltham, MA: Chronica Botanica Company.

Imeson, A. P. 1992. Natural plant exudates. In *Thickening and gelling agents for food*, ed. A. P. Imeson. London: Blackie Academic & Professional.

Knox, J. P. 1995. The extracellular matrix in higher plants. 4. Developmentally regulated proteoglycans and glycoproteins of the plant cell surface. *Fed. Am. Soc. Exp. Biol. J.* 9:1004–12.

Mantell, C. L. 1947. *The water-soluble gums.* New York: Reinhold Publ. Corp.

Nazareth, M. R., Kennedy, C. E., and V. N. Bhatia. 1961. Studies on larch arabogalactan II. *J. Pharm. Sci.* 50:564–7

Nussinovitch, A. 1997. *Hydrocolloid applications. Gum technology in the food and other industries.* London: Blackie Academic & Professional.

Nussinovitch, A. 2003. *Water soluble polymer applications in foods.* Oxford, UK: Blackwell Publishing.

Nussinovitch, A. 2010. *Polymer macro- and micro-gel beads: Fundamentals and applications.* New York: Springer.

Nussinovitch, A. 2010. *Plant gum exudates of the world sources, distribution, properties, and applications.* Boca Raton: CRC Press, Taylor and Francis Group.

Nussinovitch, A., and M. Hirashima. 2014. *Cooking innovations, using hydrocolloids for thickening, gelling and emulsification.* Boca Raton: CRC Press, Taylor & Francis Group.

Ponder, G. R., and G. N. Richards. 1997. Arabinogalactan from western larch. III: alkaline degradation revisited, with novel conclusions on molecular structure. *Carbohydr. Polym.* 34:251–61.

Stanko, G. L. 1966. Artificial sweetener-arabinogalactan composition and edible food-stuff utilizing same. U.S. Patent No. 3,294,544.

Stephen, A. M. 1983. Other plant polysaccharides. In *The polysaccharides*, vol. 2, ed. G. O. Aspinall, 97–193. Orlando: Academic Press.

Stout, A. W. 1959. Larch arabinogalactan. In *Industrial gums*, ed. R. L. Whistler, 307–10, New York: Academic Press.

Timell, T. E. 1978. Wood hemicelluloses: Part II. In *Advances in botanical research*, ed. H. W. Woolhouse, 410–33. London: Academic Press.

White, E. V. 1942. The constitution of arabogalactan. *J. Am. Chem. Soc.* 64:2838–42.
Wise, L. E., Peterson, F. C., Barry, A. J., and H. Unkauf. 1940. The chemistry of wood. VII. The esters and ethers of the water-soluble polysaccharides of larch wood. *J. Am. Chem. Soc.* 62:2361–5.

CHAPTER 12

Levan

12.1 INTRODUCTION

The most important microbial polysaccharides—xanthan gum, curdlan, gellan gum, and bacterial cellulose—and their recipes were described in our previous book: *Cooking Innovations, Using Hydrocolloids for Thickening, Gelling, and Emulsification*. Additional microbial polysaccharides can have potential applications in foods, and the current chapter aims to provide the relevant details related to levan's manufacture and structure, production, functionality, and food applications.

12.2 MANUFACTURE AND STRUCTURE

Levan is an extracellular polysaccharide. It can be produced by both a variety of microorganisms, such as *Bacillus subtilis*, *Bacillus polymyxa*, *Aerobacter levanicum*, *Streptococcus* sp., *Pseudomonas* sp., and *Corynebacterium laevaniformans*, and by plants.

When the gum is produced by *B. polymyxa*, the medium is composed of sucrose, peptone, yeast extract, K_2HPO_4, $(NH_4)_2SO_4$, and $MgSO_4$.

A small amount of levan can be obtained when lactose, maltose, and raffinose are used as carbon sources. When glucose or fructose is used as the carbon source, no levan is produced. Cultivation for at least 10 days is needed for maximal yield. During production, a decrease in pH from 7.0 to 4.7 occurs due to the acid produced. Optimal temperature for the reaction is ~30°C. Gum production is enhanced by gentle shaking of the culture, but vigorous agitation and aeration decrease yields. After producing the gum, it is centrifuged to remove the bacterial cells. The gum is harvested by precipitation in ethanol or isopropanol. The product is freeze-dried or vacuum-dried and can be further purified. Levan is composed of repeating 5-member fructofuranosyl rings connected by β-2,6 linkages. The main chain is branched through β-2,1 linkages of the fructofuranosyl rings. Both molecular weight and degree of branching are influenced by the type of organism producing the gum. The molecular weight of bacterial levans ($2–100 \times 10^6$) is higher than that of plant levans ($2–33 \times 10^3$).

12.3 PROPERTIES

Levan does not swell in water and has very low intrinsic viscosity, i.e., 0.07–0.18 dL/g, at molecular weights of $16–24 \times 10^6$. The gum has an ellipsoidal molecular shape. Flexible and cohesive films can be produced from levan by the addition of 20%–30% glycerol. At a glycerol concentration of <10%, more brittle films are produced, whereas at concentrations of >30%, glycerol films are not cohesive.

12.4 FOOD APPLICATIONS

12.4.1 General Approach

Levan is naturally present in several food products, and consequently, it is consumed in minute quantities by humans. Nevertheless, until recently, it was ignored as a functional food constituent owing to its limited resources and very low concentration in food. In the food industry, levan is used as a stabilizer, emulsifier, formulation aid, surface-finishing agent, encapsulating agent, and a carrier of flavors and fragrances. In addition, the enzymatic or chemical hydrolytic products of levan may be used in the food industry as sweeteners or dietary fiber, for example β-2,6-linked fructofuranosyl oligosaccharides.

12.4.2 Prebiotic Effects

Levan from *Lactobacillus sanfranciscensis*, LTH 2590, displays prebiotic effects, and it has attracted attention for its antitumor and cholesterol-lowering properties as well as for its application as an eco-friendly adhesive. Prebiotic substances stimulate the proliferation of useful bacteria in the colon while inhibiting harmful microorganisms. They are not digested until they reach the large intestine where they produce short-chain fatty acids. It has been proposed that a valuable prebiotic might contain both high- and low-molecular-weight fructans (levans and inulins). The smaller molecules would be used straightaway, whereas the larger molecules would be hydrolyzed slowly, thus prolonging the prebiotic effect.

12.4.3 Beverages and Colloid Systems

Levan can be used in beverages as a viscosity former. Even if the processing method includes heating or if the levan solution contains sodium chloride, the viscosity of the liquid at room temperature remains relatively stable. The viscosity might be influenced by acidic conditions (pH 2.0), but it is stable in the range of pH 4.0–10.0. Another study reported that levan can stabilize a colloid system by the formation of an emulsion, aerosol, or foam, which can be used in food and beverages. The enhanced colloidal stability was hypothesized to result from the interaction with proteins, but the mechanism is, in fact, not known.

12.4.4 Edible Coatings and Films

Levan is beneficial in terms of its water-holding capability. At >1% (w/v) concentration, a levan solution will form films on a suitable smooth surface. A low-concentration levan solution (<1%) can similarly be used in coatings. Levan has also been proposed as an ingredient for creating edible films. These edible films are too brittle when levan is used as the only ingredient, but when levan is mixed with other ingredients, such as glycerol, elastic and extrudable films can be produced. Glycerol is an effective plasticizer in the solid state and it reduces the glass transition temperature without undergoing degradation, thereby enabling efficient processing at low temperatures. A critical concentration of 20% glycerol is essential for effective plasticization of levan. Extrusion can incorporate glycerol into the levan structure more thoroughly, because of increased hydrogen-bonding as compared to cold-mixed and compression-molded samples. Thermal processing of levan by conventional molding and extrusion techniques is more effective. Assuming that levan can be produced cost-effectively through a microbial process, the molecule has the potential to be turned into films and molded products for applications in commodity polymers. Films made from levan when blended with chitosan and polyethylene oxide (PEO) have also been cast. Although chitosan, PEO, and levan have film-forming capabilities, they each have deficiencies as well. By changing the ratios of the components, films could be created with improved flexibility, stability, and transparency. Each of these polymers contributed to the general success: Levan and chitosan improved the mechanical and thermal stability of PEO, and levan enhanced biocompatibility.

12.5 RECIPES

12.5.1 Walnut Meringue (Figure 12.1)

(Makes 8 balls, 5 cm in diameter)

50 g egg white

2.5 g (½ tsp) 1% levan solution (Figure 12.1A, B) (Hint 12.5.1.1)

25 g (2 tbsp) granulated sugar

0.5 g (1/12 tsp) salt

30 g walnuts (Hint 12.5.1.2)

Walnut halves (optional)

1. Crush walnuts with a rolling pin.
2. Preheat oven to 120°C.

Figure 12.1 Walnut meringue. (A) Levan sample. (B) 1% levan solution. (C) Whisked egg white with levan, sugar, and salt. (D) Whisked egg white and walnuts on a baking tray before baking. (E) Baked meringue.

3. Put egg white and 1% levan solution in a bowl, and whisk for 1 minute. Then add granulated sugar and salt, and whisk for 1 minute (Figure 12.1C).
4. Add crushed walnuts (1) to whisked egg white (3).
5. Place the mixture (4) on the baking tray with a spoon (Figure 12.1D) and top with walnut halves (Hint 12.5.1.3).
6. Bake the meringues (5) in an oven at 120°C for 30 minutes, and then cool them on a wire rack (Figure 12.1E).

pH of 1% levan solution = 6.5

pH of meringue = 7.3

Preparation hints:

12.5.1.1 Levan dissolves in water easily with stirring.

12.5.1.2 Any nut can be used. Walnuts and coconut are recommended.

12.5.1.3 Placing walnut halves on top of the meringue changes the meringue's texture.

Adding levan to egg whites leads to a stiffer whisked mixture.

12.6 TIPS FOR THE AMATEUR COOK AND PROFESSIONAL CHEF

- Levans are possible substitutes for dextrans, when the gum needs to have low viscosity, high water solubility, and susceptibility to acid hydrolysis.

References and Further Reading

Arrizon, J., Hernández-Moedano, A., Toksoy Öner, E., and M. González-Avila. 2014. *In vitro* prebiotic activity of fructans with different fructosyl linkage for symbiotics elaboration. *Int. J. Probiotics* 9:69–76.

Barone, J. R., and M. Medynets. 2007. Thermally processed levan polymers. *J. Food Eng.* 83:521–30.

Bostan, M. S., Mutlu, E. C., Kazak, H., Keskin, S. S., Toksoy Öner, E., and M. S. Eroglu. 2014. Comprehensive characterization of chitosan/PEO/levan ternary blend films. *Carbohydr. Polym.* 102:993–1000.

Cote, G. L. 1992. Low-viscosity α-D-glucan fractions derived from sucrose which are resistant to enzymatic digestion. *Carbohydr. Polym.* 19:249–52.

De Vuyst, L., De Vin, F., Vaningelgem, F., and B. Degeest. 2001. Recent developments in the biosynthesis and applications of heteropolysaccharides from lactic acid bacteria. *Int. Dairy J.* 11:687–707.

Fujiwara, K., Oosawa, T., and H. Saeki. 1998. Improved thermal stability and emulsifying properties of carp myofibrillar proteins by conjugation with dextran. *J. Agric. Food Chem.* 46:1257–61.

Ghaly, A. E., Arab, F., Mahmoud, N. S., and J. Higgins. 2007. Production of levan by *Bacillus licheniformis* for use as a soil sealant in earthen manure storage structures. *Am. J. Biotechnol. Biochem.* 3:47–54.

Gounga, M. E., Xu, S.-Y., and Z. Wang. 2007. Whey protein isolate-based edible films as affected by protein concentration, glycerol ratio and pullulan addition in film formation. *J. Food Eng.* 83:521–30.

Han, Y. W. 1990. Microbial levan. *Adv. Appl. Microbiol.* 35:171–94.

Han, Y. W., and M. A. Clarke. 1990. Production and characterization of microbial levan. *J. Agric. Food Chem.* 38:393–6.

Kasapis, S., Morris, E. R., Gross, M., and K. Rudolph. 1994. Solution properties of levanpolysaccharide from *Pseudomonas syringae* pv. *phaseolicola*, and its possibleprimary role as a blocker of recognition during pathogenesis. *Carbohydr. Polym.* 23:55–64.

Khan, T., Park, J. K., and J.-H. Kwon. 2007. Functional biopolymers produced by biochemical technology considering applications in food engineering. *Korean J. Chem. Eng.* 24:816–26.

Kim, C. H., Song, K. B., and S. K. Rhee. 1998. Viscosity of levan produced by levansucrase from *Zymomonas mobilis*. *Food Eng. Progr.* 2:217–22.

Korakli, M., Pavlovic, M., Ganzle, M. G., and R. F. Vogel. 2003. Exopolysaccharide and kestose production by *Lactobacillus sanfranciscensis* LTH2590. *Appl. Environ. Microbiol.* 69:2073–9.

Park, J. K., and T. Khan. 2009. Other microbial polysaccharides: pullulan, scleroglucan, elsinan, levan, alternan, dextran. In *Handbook of hydrocolloids*, eds. G. O. Phillips and P. A. Williams, 592–615. Boca Raton: CRC and Oxford: Woodhead Publishing Limited.

Patel, A., and J. B. Prajapati. 2013. Food and health applications of exopolysaccharides producedby lactic acid bacteria. *Adv. Dairy Res.* 1:1–7.

Toksoy Öner, E., Hernández, L., and J. Combie. 2016. Review of levan polysaccharide: from a century of past experiences to future prospects. *Biotechnol. Adv.* 34:827–44.

Yoo, S.-H., Yoon, E. J., Cha, J., and H. G. Lee. 2004. Antitumor activity of levanpolysaccharides from selected microorganisms. *Int. J. Biol. Macromol.* 34:37–41.

CHAPTER 13

Mesquite Gum

13.1 INTRODUCTION

The bark of *Prosopis* spp. yields an exudate, which is identified as mesquite gum. This exudation occurs in response to insect attack, wounds or physiological stresses such as heat or severe lack of water. Mesquite gum is exuded by tree bark in the form of glassy, red- to amber-colored lumps. Mesquite trees are leguminous plants that are prevalent in dry and semiarid regions of the world. Actually, the genus *Prosopis* includes roughly forty-four dissimilar species that grow habitually in North and South America, and similarly in Africa, Australia, and eastern Asia. Today, mesquite gum is not used in household applications too often. Conversely, in historical times, mesquite gum was used extensively in food applications, and it is traditionally considered as a substitute or adulterant of gum arabic, but of inferior quality due to its darker color.

13.2 EXUDATE COMMON NAMES AND DISTRIBUTIONAL RANGE

The common names of mesquite gum are: ironwood, mesquite, *bayarone* [French], *mesquitebaum* [German], *algaroba* [Portuguese (Brazil)], *prosópis* [Portuguese (Brazil)], *algarroba* [Spanish], *algarrobo* [Spanish], and *cují negro* [Spanish]. Its distributional range covers Mexico, Costa Rica, El Salvador, Guatemala, Honduras, Nicaragua, Panama, Venezuela, Colombia, Ecuador (including The Galapagos), Peru, tropical Africa, Asia, Australia, West Indies, Mascarene Islands, and Hawaii. The mesquite taxon is *Prosopis juliflora* (Sw.) DC. and its synonyms are *Mimosa juliflora* Sw. [≡ *Prosopis juliflora* var. *juliflora*]; *Prosopis horrida* Kunth [≡ *Prosopis juliflora* var. *horrida*]; *Prosopis vidaliana* Náves [*Prosopis juliflora* var. *juliflora*].

13.3 EXUDATE APPEARANCE

The gum of *P. juliflora* exudes spontaneously from the stems and branches during the summer months. The gum exudates in a semi-fluid, sticky, soft condition and then hardens in only a few hours, forming tear drops of various sizes and colors; these gum drops whiten with exposure to sunlight—ultimately becoming translucent—and are frequently filled with small fissures. The resultant exudate is incredibly brittle. In the dry summers, some gum will be found on the ground beneath the tree. There are numerous regions where the gum is less frequently encountered due to changing weather conditions. The plant heals the scars generated by tapping and regularly secretes gum along the edges of the old wounds. The gum is gathered manually from small trees. It has irritant properties, and reportedly its consumption over long periods of time results in the death of cattle. Gum of *P. juliflora* has been reported to exude in unevenly round or vermiform pieces weighing 5–25 g each. The gum of *P. juliflora* and other varieties found in similar regions is usually clear and yellowish or light amber to brown, but it can also be markedly red and it darkens with age.

13.4 GUM WATER SOLUBILITY

This gum, when of normal quality, is not always entirely soluble in water. The fraction that is not readily soluble swells and forms a soft jelly. The gum dissolves completely in its equivalent weight of water in 24 hours, at a temperature of ~21°C. Its solubility is comparable to that of the medium or poorer grades of Sudan gum arabic.

13.5 GUM CHEMICAL CHARACTERISTICS

Analytical and structural studies have been performed on *Prosopis* gums. The gum of *P. juliflora* is the natural salt of a complex acidic polysaccharide. It has 11% moisture, 0.7% protein, and 2%–4% ash. A minute quantity of protein may result from the enzyme that is responsible for the formation of the gum, or from contact of the gum with proteinaceous material in the tree. The composition of the ash indicates that the gum is mostly a calcium salt. It contains L-arabinose, D-galactose, and 4-O-methyl-D-glucuronic acid, identified as the α and β anomers of methyl (4-O-methyl-D-glucoside) uronamide. These three constituents are present in a molar ratio of 4:2:1. Several structures have been suggested and one of them has three D-galactopyranose units in the main chain united by (1,3) linkages, and there is also a side chain attached to the C6 in each of these units. As in the case of gum arabic, it is also likely that the three D-galactose units are joined by (1,6) or (1,3) and (1,6) linkages. The free acid form has an equivalent molecular weight of 1,350 and contains 2.9% methoxyl groups.

13.6 COMMERCIAL AVAILABILITY OF THE GUM AND APPLICATIONS

From the 1940s–1960s, this gum was gathered and marketed in Mexico, South America, and the southwestern United States. The relatively low viscosity of the aqueous solutions of *Prosopis* gum made it a practical substitute for gum talha and the technological grades of gum arabic. At the end of the 19th century, an average annual yield of 5,400 kg was reported from Texas alone. In general, it is not readily available now, most likely due to its eradication from areas in which mesquite was considered as a thorny pest. Nevertheless, this gum may be used for agroforestry in areas where windbreaks or soil stabilization and enrichment are needed. A straightforward and practical technological process, based on the extraction of water-soluble constituents from mesquite pods (*Prosopis pallida*), can be used to obtain good quality syrups and dietary fiber concentrates that can be used as food ingredients. Because of the high sugar (up to 500 g/kg) and dietary fiber (322.2 g/kg) content of these mesquite pods, they can be used in agroindustries, and this enhances their economic value. In the past, the gum has been used as a binder in tablet dosage forms, as an emulsifying agent to encapsulate citrus essential oils, to relieve sore throat and irritated eyes, and as an antidote to lice. In the latter application, the boiled gum was often mixed with mud and plastered on the hair for a day or two. When the "pack" was removed, the hair was dyed black, and was glossy and free of lice.

Due to the presence of tannins, mesquite gum is not permitted as a food additive in the U.S. However, it is used in domestic cooking in the Sonora region of northwestern Mexico, to prepare a traditional dessert known as *capirotada*. Today it is mainly used in the ink, textile, and glue industries. It is also used as an emulsifier in confectioneries (where approved) and for mending pottery. The Apache chewed the gum as a candy, and it is surprisingly sweet when burned. Mesquite gum has been used as a raw material for the preparation of L-arabinose, as reported in some of the standard chemical methods. Other gums proposed for the same uses are cherry, peach, and Australian black wattle gums. The production and use of edible films formed from biopolymers are on the rise. Films prepared from mesquite gum as a structural agent can be used to extend the shelf life of fruits and vegetables. The type and concentration of the dispersed hydrophobic phase affect the mean droplet size and the distribution of the emulsion. The smaller these two parameters are, the more homogeneous and unflawed the film morphology is likely to be, and the lower is its water vapor permeability. Mesquite reportedly has the potential to be used in spray-drying applications. Cardamom-based oil microcapsules were successfully produced by spray-drying with mesquite gum. The stability of emulsions against drop coalescence was increased for all gum-to-oil ratios that were studied. High flavor retention (83.6%) was attained during microencapsulation by spray-drying when a gum-to-oil proportion of 4:1 was used. This confirmed the interesting emulsifying properties and the good flavor-encapsulation abilities of mesquite gum, which qualifies it as an important alternative encapsulating medium. The microcapsules can be readily used as a food ingredient. Blueberry is an important source of anthocyanins, which are strongly colored substances known for their antioxidant activity. One of the drawbacks of using anthocyanins as a food colorant is their low stability. Some spray-dried powders were obtained from blueberry extracts with added mesquite gum; variations in the color and concentration of the compounds (which produce the color) present in these powders were evaluated. Mesquite gum served as a good protective agent for the color because it reduced anthocyanin degradation in the microencapsulates that were exposed to light and temperatures of 4°C and 25°C. After 4 weeks of storage at 4°C in the dark, the sample demonstrated negligible changes in color, showing that this process is advantageous for the conservation of colorants.

13.7 RECIPES

13.7.1 Cooking with Mesquite Meal

Mesquite meal is prepared by manual stone grinding of the seeds and pods. Recent milling methods have accelerated the process by grinding the entire mesquite pod, including its protein-rich seed. This produces a highly nutritious and tasty meal. It contains up to 17% protein and has high lysine content. Thus, it can serve as a perfect addition to other grains that are low in this amino acid. The seed mucilage or seed coat is included in the ground meal. The galactose-to-mannose ratio and the food applications of the gum are similar to those of guar gum. Even though desert dwellers have used mesquite pods as a source of food for centuries, amateur cooks, chefs, and nutritionists who experiment with the uses of mesquite meal are the ones who brought this tasty ingredient into the modern kitchens.

13.7.2 Cornbread (Figure 13.1)

(Makes 21 cm diameter loaf)
55 g (¼ cup) butter
250 g (1 cup) milk
1 egg
210 g (1¼ cup) cornmeal
140 g (1 cup) all-purpose flour
2.0 g (a little under 1 tsp) mesquite powder

Figure 13.1 Cornbread. (A) Melt butter on low heat. (B) Mix melted butter, milk, and egg. (C) Add cornmeal, all-purpose flour, granulated sugar, mesquite powder, baking powder, and salt. (D) Mix all ingredients. (E) Pour batter into a round pan of 21 cm. (F) Bake for 25 minutes at 200°C. (G) Serve.

100 g (a little under ½ cup) granulated sugar
12 g (1 tbsp) baking powder
3 g (½ tsp) salt
Vegetable oil

1. Preheat oven to 200°C.
2. Spread vegetable oil on the bottom and sides of a round pan having a diameter of 21 cm.
3. Put butter in a small saucepan and heat on low until it melts (Figure 13.1A).
4. Put the melted butter (3), milk (Hint 13.7.2.1), and egg in a bowl and mix with a whisk (Figure 13.1B).
5. Add cornmeal, all-purpose flour, granulated sugar, mesquite powder, baking powder, and salt to the mixture (4) (Figure 13.1C) and stir with a spatula (Figure 13.1D).
6. Pour the batter (5) into the round pan (2) (Figure 13.1E) and bake for 25 minutes at 200°C (Figure 13.1F) (Hint 13.7.2.2).

pH of dough = 6.3

Preparation hints:

13.7.2.1 When cold milk is added to the butter, the butter will partially solidify. This is fine.

13.7.2.2 This bread is best served warm (Figure 13.1G).

Adding mesquite powder to cornbread increases its savory flavor like cocoa, but adding too much mesquite powder results in a crumbly and brittle texture. Less than 1.0 wt% of the flour and cornmeal is best.

13.7.3 Healthy Cornbread (Figure 13.2)

(Makes 15 cm diameter loaf)

140 g (¾ cup) cornmeal
100 g (¾ cup) strong flour
35 g (⅜ cup) mesquite power
2 g (½ tsp) baking soda

FIGURE 13.2 HEALTHY CORNBREAD. (A) MIX DRY INGREDIENTS. (B) MIX LIQUID INGREDIENTS IN A SEPARATE BOWL. (C) ADD LIQUID MIXTURE TO DRY MIXTURE. (D) MIX ALL INGREDIENTS WELL. (E) BAKE FOR 20 MINUTES AT 180°C. (F) SERVE.

3g (½ tsp) salt
240g (1 cup) yogurt
1 egg
60g (3 tbsp) honey
36g (3 tbsp) vegetable oil

1. Preheat oven to 180°C.
2. Spread vegetable oil on the bottom and sides of a round pan having a diameter of 15cm.
3. Put cornmeal, strong flour, mesquite powder, baking soda, and salt in a bowl and mix with a spatula (Figure 13.2A).
4. Put yogurt, egg, honey, and vegetable oil in another bowl (Figure 13.2B) and mix with a whisk (Hint 13.7.3.1).
5. Add liquid mixture (4) to dry mixture (3) (Figure 13.2C) and stir with a spatula (Figure 13.2D).
6. Pour the batter (5) into the pan (2) and bake for 20 minutes at 180°C (Figure 13.1E) (Hint 13.7.3.2).

pH of dough = 5.9

Preparation hints:

13.7.3.1 Lecithin within the egg yolk is an emulsifier which will emulsify the yogurt and vegetable oil, so mix strongly with a whisk to avoid separation.

13.7.3.2 This bread is best served warm (Figure 13.2F).

This cornbread has a lower lipid content and smoother texture than the regular cornbread (Figure 13.1). It also has a savory flavor like cocoa.

13.8 TIPS FOR THE AMATEUR COOK AND PROFESSIONAL CHEF

- Mesquite gum solutions are effective in the preparation and stabilization of oil-in-water emulsions.
- Addition of mesquite gum in the range of 0.8%–1.2% (w/w) to bread formulations improves their sensory evaluation and extends their shelf life.

References and Further Reading

American Pharmaceutical Association. 1946. *The committee on national formulary*, 8th edn., 67–8. Washington DC: American Pharmaceutical Association.

Anderson, D. M. W., and J. G. K. Farquhar. 1982. Gum exudates from the genus *Prosopis* (leguminous tree, gum mesquite). *Int. Tree Crops J.* 2:15–24.

Balls, E. K. 1962. *Early uses of California plants*. Berkeley: The University of California Press.

Ben-Zion, O., and A. Nussinovitch. 1997. Physical properties of hydrocolloid wet glues. *Food Hydrocolloids* 11:429–42.

Beristain, C. I., Garcia, H. S., and E. J. Vernon-Carter. 2001. Spray-dried encapsulation of cardamom (*Elettaria cardamomum*) essential oil with mesquite (*Prosopis juliflora*) gum. *LWT - Food Sci. Technol.* 34:398–401.

Bravo, L., Grados, N., and F. Saura-Calixto. 1998. Characterization of syrups and dietary fiber obtained from mesquite pods (*Prosopis pallida* L). *J. Agric. Food Chem.* 46:1727–33.

Churms, S. C., Merrifield, E. H., and A. M. Stephen. 1978. Regularity within the molecular structure of arabinogalactan from Western larch (*Larix occidentalis*). *Carbohydr. Res.* 64:C1–2.

Diaz-Sobac, R., Garcia, H., Beristain, C. I., and E. J. Vernon-Carte. 2002. Morphology and water vapour permeability of emulsion films based on mesquite gum. *J. Food Process. Preserv.* 26:129–41.

Edwards, H. G. M., Falk, M. J., Sibley, M. G., Alvarez, J. B., and F. Rull. 1998. FT-Raman spectroscopy of gums of technological significance. *Spectrochimica Acta, Part A—Mol. Biomol. Spectroscopy* 54A:903–13.

Felker, P., and R. S. Bandurski. 1979. Uses and potential uses of leguminous trees for minimal energy input agriculture. *Econ. Bot.* 33:172–84.

Felter, H. W., and J. U. Lloyd. 1898. *King's American dispensatory*, 18th edn., 3rd revision, reprinted 1983. Portland, OR: Eclectic Medical Publications.

Ferri, C. M. 1959. Gum tragacanth. In *Industrial gums*, ed. R. L. Whistler, 511–5. New York: Academic Press.

Foreb, R. H. 1895. The mesquite tree: its properties and uses. *Arizona Agric. Exp. Station Bull.* 13:15–26.

Fowells, H. A. 1965. U.S. Department of Agriculture, Forest Service, Silvis of forest trees of the United States. Washington DC: Agriculture Handbook 271.

Glicksman, M. 1963. Natural gums. In *Kirk-Othmer encyclopedia of chemical technology*, 2nd edn., vol. 10, 741–54. New York: Wiley.

Glicksman, M. 1983. Gum karaya (*Sterculia gum*) [Manufacture, structure, properties, applications in pharmaceuticals and foods, regulatory status]. *Food Hydrocolloids* 2:39–47.

Glicksman, M. 1983. Gum tragacanth: Regulatory status, structure, properties, food applications (*Astragalus gummifer*). *Food Hydrocolloids* 2:49–60.

Goldstein, A. M., and E. N. Alter. 1959. Gum karaya. In *Industrial gums*, ed. R. L. Whistler, 343–59. New York: Academic Press.

Goycoolea, F. M., Calderon de la Barca, A. M., Balderrama, J. R., and J. R. Valenzuela. 1997. Immunological and functional properties of the exudate gum from northwestern Mexican mesquite (*Prosopis* spp.) in comparison with gum arabic. *Int. J. Biol. Macromol.* 21:29–36.

Goycoolea, F. M., Morris, E. R., Richardson, R. K., and A. E. Bell. 1995. Solution rheology of mesquite gum in comparison with gum Arabic. *Carbohydr. Polym.* 27:37–45.

Howes, F. N. 1949. *Vegetable gums and resins*. Waltham, MA: Chronica Botanica Company.

Imeson, A. P. 1992. Natural plant exudates. In *Thickening and gelling agents for food*, ed. A. P. Imeson. London: Blackie Academic & Professional.

Jimenez-Aguilar, D. M., Ortega-Regules, A. E., Lozada-Ramirez, J. D., Perez-Perez, M. C. I., Vernon-Carter, E. J., and J. Welti-Chanes. 2011. Color and chemical stability of spray-dried blueberry extract using mesquite gum as wall material. *J. Food Composition Anal.* 24:889–94.

Kesar, S. 1935. Kullu (*Sterculia urens*). Wailings from Damoh, C. P. *Indian Forester* 61:144–9.

Kubal, J. V., and N. Gralen. 1948. Physico-chemical properties of karaya gum. *J. Colloid Sci.* 3:457–71.

Lewis, W. H., and M. P. F. Elvin-Lewis. 1977. *Medical botany*. New York: John Wiley & Sons.

Mantell, C. L. 1947. *The water-soluble gums*. New York: Reinhold Publ. Corp.

Nussinovitch, A. 1997. *Hydrocolloid applications. Gum technology in the food and other industries*. London: Blackie Academic & Professional.

Nussinovitch, A. 2003. *Water soluble polymer applications in foods*. Oxford, UK: Blackwell Publishing.

Nussinovitch, A. 2010. *Polymer macro- and micro-gel beads: Fundamentals and applications*. New York: Springer.

Nussinovitch, A. 2010. *Plant gum exudates of the world sources, distribution, properties, and applications*. Boca Raton: CRC Press, Taylor and Francis Group.

Nussinovitch, A., and M. Hirashima. 2014. *Cooking innovations using hydrocolloids for thickening, gelling, and emulsification*. Boca Raton: CRC Press, Taylor and Francis Group.

Sybil, P. J., and F. Smith. 1945. The chemistry of gum tragacanth. Part III. *J. Chem. Soc.* 1945:749–51.

Timell, T. E. 1978. Wood hemicelluloses: Part II. In *Advances in botanical research*, ed. H. W. Woolhouse, 410–33. London: Academic Press.

USDA, ARS, National Genetic Resources Program. 2008. *Germplasm resources information network (GRIN)*. [Online Database] National Germplasm Resources Laboratory, Beltsville, MD. Available at: http://www.ars-grin.gov/cgi-bin/npgs/acc/display.pl.

Verma, V. P. S., and G. N. Kharakwal. 1977. Experimental tapping of *Sterculia villosa* Roxb. for gum karaya. *Indian Forester* 103:269–72.

Weiping, W., and A. Branwell. 2000. Tragacanth and karaya. In *Handbook of hydrocolloids*, eds. G. O. Philips and P. A. Williams, 231–46. Cambridge: Woodhead Publishing Ltd.

Whistler, R. L., and C. L. Smart. 1953. Gum Arabic. In *Polysaccharide chemistry*, 319–26. New York: Academic Press.

White, C. S. 1951. Improvements in the preparation of L-arabinose from Mesquite gum. *J. Am. Chem. Soc.* 73:4038–9.

White, E. V. 1946. The constitution of mesquite gum. I. The methanolysis products of methylated mesquite gum. *J. Am. Chem. Soc.* 68:272–5.
White, E. V. 1947. The constitution of mesquite gum. II. Partial hydrolysis of mesquite gum. *J. Am. Chem. Soc.* 69:622–3.

CHAPTER 14

Milk Proteins

14.1 INTRODUCTION

Protein-enriched ingredients have high nutritional value. Such constituents can bind water and fat, entrap water and gas, and emulsify fat. These abilities serve to create a desired texture and, of course, they influence sensory evaluation. Both bioactive peptides and hydrolysates from milk protein can play potentially important roles in different food items.

14.2 THE MILK PROTEIN SYSTEM

Bovine milk has 30–35.8 g/L proteins in total. The main components of these proteins are caseins and whey proteins. Caseins include α_{S1}-casein, β-casein, κ-casein, and α_{S2}-casein. Whey proteins include β-lactoglobulin, α-lactalbumin, immunoglobulins, bovine serum albumin (BSA), and proteose peptones. Milk also contains 37–50 g/L fat, 47–49.6 g/L lactose, 6.8–7.4 g/L ash, and 845–877 g/L water. Thus, milk is a complex system that contains

~3.5% protein by weight. Caseins account for ~80% of the nitrogen in milk. They are insoluble at ~pH 4.6 (their isoelectric point) and at temperatures >8°C, and they precipitate under these conditions. As noted, the casein fraction is made up of four main proteins and other minor proteins and peptides. α_{S1}-casein and β-casein do not contain cysteine or cystine, whereas κ-casein and α_{S2}-casein have two half-cystine residues. Caseins can associate due to the presence of high hydrophobicity regions, charge allocation, phosphorylation, and glycosylation.

Milk caseins are ~30–600 nm in size. Their molecular mass is in the range of 10^5 kDa. They can be separated from whey protein by ultracentrifugation. Caseins, as micelles (more or less spherically-shaped colloidal associations), are present at concentrations of 10^{14}–10^{16} per mL of milk; these micelles are situated 2 micelle diameters apart. The micelles also contain citrate, potassium, sodium, magnesium, and colloidal calcium phosphate. The micelle is stabilized by κ-casein, present mostly at the micelle surface. In raw milk, the aggregation of micelles is prevented by both electrostatic and steric repulsions owing to the "hairy coat" layer of the κ-casein. Removal of this coat destabilizes the micelle and, thus, calcium-mediated aggregation can take place depending upon pH, temperature, and ionic environment. The whey protein fraction is quite heterogeneous and the whey proteins have a globular conformation. A considerable segment of their sequence is structurally ordered. These structures are sensitive to heat, but are less sensitive to changes in ionic strength and pH than caseins. Whey proteins get denatured upon heating at ~70°C. At 72°C, they denature and interact with casein to form a complex.

14.3 MILK PROTEIN PRODUCTS

Powdered dry milk protein products have a long shelf life and can be used in numerous food applications. Casein is produced by lowering the pH of milk to ~4.6, by means of acidification via different chemical or biological methods. Another approach is to destabilize the casein micelles at the natural pH of milk by using proteolytic enzymes. The modified casein micelle becomes susceptible to the calcium present in the serum phase of the milk, and coagulates at temperatures >20°C. Increasing the temperature to ~50°C–60°C causes further coagulation and then the formed curd can be separated by various methods or devices. The next step is to wash the curd to remove lactose, salts, and whey proteins, followed by mechanical dewatering and drying. To obtain uniform moisture, the dried powder is tempered, blended, and ground to the required particle size. There are various methods of producing caseinates that differ in their properties and applications. The micellar casein is compositionally similar to the native casein micelles found in milk.

Spray-dried sodium caseinate is water-soluble and is mostly used in food applications. Acid caseins are not soluble upon redispersion in water, but can be solubilized by the addition of alkali. Details on the production of different caseins and caseinates, and on casein fractionation can be found elsewhere.

Milk whey is the serum or liquid that remains once the fat and casein are removed for the production of cheese or acid and rennet casein. The industrial procedures used to isolate protein products from whey are summarized in many sources. Products include whey powders and modified whey powders, whey protein concentrate (WPC), whey protein isolate (WPI), and lactalbumin. Whey proteins can be coprecipitated with casein by heating milk at its natural pH, or by inducing denaturation that is followed by complexation with casein and precipitation of the complex via acidification or addition of $CaCl_2$. Milk protein concentrate (MPC) is a high-protein spray-dried powdery product manufactured from skimmed milk. In MPC, the casein has a micellar form comparable to that found in milk, while the whey proteins are in their native form. In addition, such products have a relatively high ash content since the protein-bound minerals are retained. This content can be reduced by acidification. Modified milk protein products are needed due to their functional properties in specific food products. These consumables can be produced by changing the physicochemical properties of milk proteins. Milk protein hydrolysates and biologically-active peptide fractions are also produced for specific functional and nutritional applications.

14.4 FUNCTIONAL PROPERTIES OF MILK PROTEIN PRODUCTS

Both chemical and physical properties of milk proteins will influence their functionality in different food products. A most important property is their solubility, which is influenced by several parameters, such as hydrophobic interactions, intermolecular electrostatic interactions, and their balance. The insoluble acid casein (in water at pH 4.6) can be converted to caseinate (i.e. completely soluble cationic salt at pH > 5.5). Caseinates in high concentrations are soluble on both the acidic and alkaline sides of their isoelectric point. Attention should also be given to a possible increase in viscosity at high concentration. Substantial differences in solubility can be observed among the commercial MPC products that have different protein contents. Nevertheless, they can be dissolved at 50°C. MPC solubility can be improved by adding 0.035–0.25 M monovalent salts to the ultrafiltered retentate prior to drying. Whey proteins, in their native globular conformation, are soluble over the whole pH range at ionic strength >0.025 M. At high salt concentrations, their solubility decreases as a result of salting out.

In most cases, when milk undergoes gelation under chemical, enzymatic, or heating treatments, casein is the involved constituent. Protein gelation results from unfolding of the protein structure and subsequent interactions of the polypeptide segments to create a 3D cross-linked network that entraps water. In addition, proteolysis of milk to hydrolyze the micelle-stabilizing κ-casein, produces para-κ-casein-containing micelles that coagulate in the presence of Ca^{2+} in the milk serum. Another option for gelation is via acid. Thermal treatment can be used to produce thermal gels from whey proteins. Depending on solution conditions, different types of gels can be formed. Gels can also be produced via a process in which WPIs, at neutral pH and low ionic strength, are heated to an appropriate temperature and then cooled, followed by acidification by an acid precursor or infusion of salt into the cold solution. Weak, brittle gels will be produced by rapid acidification, whereas slower acidification results in stronger gels.

Caseinates form viscous solutions at concentrations above ~15%, as a result of their hydration and water-binding or entrapping properties.

Furthermore, hydration and swelling increase the hydrodynamic volume and polymer–polymer interactions, which in turn increase viscosity. Solutions that include <12% caseinates demonstrate Newtonian-type flow. At concentrations >12%, pseudoplastic-type flow dominates. Even when solutions contain higher concentrations of protein (e.g., >20%), viscosity might increase to a point where processing becomes difficult. The viscosity of sodium caseinate also depends on pH. Minimal viscosity is achieved at ~pH 7.0; at pH 9.0, viscosity increases and above pH 9.0, the viscosity decreases because then the protein molecules start to behave like disconnected entities. Casein viscosity is also a great deal higher at low pH (i.e., ~2.5–3.5) than at neutral pH. In the former case, gel-like structures form when protein contents are >5%, at temperatures <40°C. The viscosity also depends on the cations that are present, with calcium caseinate showing lower viscosity than sodium caseinate and, in general, Na > K > NH_4. Of course, manufacturing conditions strongly influence the viscosity of casein/caseinates, and details on this can be found elsewhere.

Milk protein products can be used for both emulsifying and foaming purposes. Soybean emulsions can be stabilized by the addition of sodium caseinate. Such emulsions present less creaming stability compared to the same emulsions stabilized with soy isolate or WPC. Caseinates produce foams that have higher overrun and lower stability as compared to foams stabilized by egg whites, WPC or WPI. Various foaming applications use whey protein-enriched products. Foaming abilities increase with increasing WPC solid content. Milk protein constituents can be modified to improve emulsifying and foaming properties. By cross-linking milk proteins, a polymerized protein is formed and this polymerization improves emulsion storage stability and forms a solid-like protein gel that envelops the emulsion droplets.

14.5 FOOD APPLICATIONS

Milk proteins can be found in various products, serving different purposes. In baked goods, a high calcium co-precipitate can affect the color, shine, nutritional value, texture, and appearance of pastry glazes, cake mixes for diabetics, milk biscuits, and cookies. Inclusion of low-calcium co-precipitate in cookies affects their nutritional value, texture, and appearance. Casein can influence the nutritional properties of breakfast cereals and high-protein breads. Calcium caseinate is used in milk biscuits and highly nutritive biscuits. Sodium caseinate is used in frozen prepared cakes, protein-enriched milk biscuits, shortening, pie filling, and powdered friable fat, as a fat-encapsulation agent and fat stabilizer. Acid casein is used in non-fat dry milk substitute to build structure in the dough and to improve its nutrition and flavor. Finally, WPC can be added to cookies, muffins, and croissants for water retention, color improvement, for added nutritional value, and to affect flavor. WPC added to breads or cakes is involved in dough formation, fat binding, and heat setting of the products.

Milk proteins can be used in dairy products also. A few examples are the use of calcium caseinate in processed cheese spread, Mozzarella cheese, and imitation cream cheese to maximize spreadability and stretch, and to serve as an emulsifier. MPC fortification and fat-to-protein standardization are useful methods, which can also be employed in small dairies to improve cheese yield and to reduce its daily variability. The yield increase is due to the higher protein retention and the slightly higher moisture content of the cheese. An increase in cheese yield is possible with curd stretching. Enhancing fat recovery can influence the amount of MPC to be added, thus improving cheese yield even more. Sodium caseinate is used in many applications. In coffee creamer, UHT cream, non-dairy creamer, imitation sour cream, and cultured cream, the added milk protein serves as an emulsifier and whitener, to achieve the desired texture, body, and sensory evaluation; the milk proteins also help in water-binding, increasing viscosity, and in amending the product's flavor. In yogurts or fruit yogurts, sodium caseinate reduces syneresis, increases gel firmness, and produces a better consistency. On one hand, sodium caseinate is used in imitation products (i.e., imitation milk, milk shakes, Mozzarella cheese, cream cheese) and in ice creams as an emulsifier, to increase foam stability and yield; and on the other hand, acid casein is used in simulated cheese, cheese analogues, and imitation milk to modify texture, meltability, and cost. When rennet casein was partially replaced with MPC for the production of spread-type processed cheese analogues, firmness and sliceability (resistance to cutting) increased, but meltability decreased.

Sensory evaluation indicated that replacement of up to 30% rennet casein by MPC leads to acceptable products, whereas replacement at, for example, 45% resulted in undesirable properties. Potassium caseinate is included in fat-reduced milk for specific nutritional purposes. WPC can be used in soft-serve ice cream and cheese products to affect the overrun and water- and fat-binding, and to increase emulsification and reduce cost. In yogurt and coffee whiteners, WPC is used to reduce lactose, and as a replacer for casein or caseinate. Finally, whey proteins are used in yogurt and Ricotta to increase yield and improve curd cohesiveness and consistency.

Milk proteins are often used to improve factors related to the esthetics of beverages. For example, casein is used in beer and wine to clarify the fluid, to remove phenolic compounds and color and, at the same time, it serves as a stabilizer. Potassium caseinate is used in white wine to remove tannins and phenolic compounds and, thus, to improve the sensory evaluation; it is also used for color removal. Sodium caseinate is used for color removal in apple juice and for prevention of "tea cream" in soluble tea products. Its effect as a stabilizer is known in chocolate beverages. Casein hydrolysate is utilized for its whipping and foaming abilities in non-alcoholic fruit beverages. WPC is applied in citrus-based beverages, soft drinks, hot cocoa beverages, and chocolate drinks to attain colloidal and foam stability and to improve nutritional aspects and flavor.

In case of dessert-type products, milk proteins, such as sodium caseinate, are used in different whipped products (i.e., whipped dessert, whipping fat, whipped topping, mousse, ice creams, frozen desserts, and puddings) and in instant desserts to affect whippability, increase overrun, for fat encapsulation, as an emulsifier and stabilizer, and for texture and flavor enhancement. Two MPCs (56% and 85%) were studied as substitutes in an ice cream mix, for 20% and 50% of the protein content. The MPCs did not modify the ice cream's physical properties significantly, on a constant protein basis, when up to 50% of the protein supplied by non-fat dry milk was replaced. MPCs may offer ice cream manufacturers an alternative source of non-fat milk solids, especially in mixes having reduced lactose or fat content, where higher concentrations of non-fat milk solids are needed to compensate for other losses in total solids. In spongy desserts, caseinate is used for whippability and aeration and to get whipped toppings, while hydrolyzed sodium caseinate is used to increase freeze-thaw stability. It is also possible to significantly alter the rheological and sensory attributes of porridges prepared from wheat grits (*dalia*), by the addition of soy protein isolate and skimmed milk powder. In addition, protein content in the resulting product is higher than that in wheat grits alone.

WPCs are used to enhance flavor in fruit jellies and jams, to increase whippability and overrun, and to reduce the cost of whipped toppings and frozen desserts or puddings. In pasta products, milk proteins are mainly added to improve the texture and nutritional aspects of the product. For example, sodium caseinate in protein-enriched pasta, calcium caseinate in wheat macaroni or high-protein pasta, casein in enriched/fortified macaroni, and WPC in pasta, all contribute to the nutritional value of the products. In most confectionary products, the inclusion of milk proteins improves the nutritional aspects of the manufactured edible items. Sometimes, the addition has special effects. For example, calcium caseinate improves the aeration of aerated cake icing; another example is the inclusion of milk protein hydrolysate in chocolate formulations to achieve relaxation of the mass. In meat products containing ground meat, inclusion of a few kinds of meat or fat to the formulation requires good emulsification, increased stability, and prevention of lipid oxidation. Sodium caseinate serves as a good binder and emulsifier in sausages.

Potassium caseinate serves as an emulsifier in low-fat meat paste. The sodium salt in the co-precipitate serves to improve the sensory evaluation of pate sausage. WPCs contribute to improved performance, water- and fat-binding, and reduced cost of various products.

In addition, milk protein products play a role in the manufacture and development of unique products. For example, sodium caseinate is used in a candy that can be stored and then eaten in space, in a special enriched dairy drink for infants, and in meat replacements. In all of these, the protein is used to enhance the nutritional value of the products. Calcium caseinate can be used as a flour substitute in baked goods for diabetics, as a meat replacement, and in infant foods for its nutritional contribution. High-calcium co-precipitate is used in cake mixes for diabetics. WPCs improve the digestibility of infant formula. WPIs in nutritional protein bars or beverages improve immune and nervous system activity. In addition, whey protein hydrolysate affects nutrition in hypoallergenic infant formulas, and casein phosphopeptide is utilized for its mineral-binding activity in dental filling materials.

Milk protein products are beneficial in different convenience food items. For example, sodium caseinate gives good emulsification and desired texture in dry cream products for sauces, in soups, nut substitutes, and imitation potato skin shells. Casein is included in synthetic caviar for better sensory evaluation, while caseinate is used in nut-like food for film formation, and in gravy mix as a whitening agent. Another important use of casein is in puffed food (i.e., textured products). Other textured items are extruded milk proteins, acid casein, fine breads, biscuits containing potassium caseinate, dietary fiber snacks containing rennet casein or co-precipitate or whey protein, and *surimi* containing WPCs that give a desired texture and reduced costs. Lastly, caseins/caseinates can be used for the formation of different edible films and coatings, and information about this can be found elsewhere.

14.6 RECIPES

14.6.1 Cottage Cheese and Whey (Figure 14.1)

(Makes 85 g cottage cheese, 430 g whey)
500 mL (2 cups) milk
50 mL (⅕ cup) vinegar or lemon juice
Dish towel or paper towel

1. Heat milk to 60°C in a microwave oven.
2. Pour vinegar into the hot milk (1) (Figure 14.1A), stir gently, and then let it stand for 10 minutes (Figure 14.1B) (Hint 14.6.1.1).
3. Strain the milk through a dish towel or paper towel (Figure 14.1C). The liquid is whey and the white solid in the dish towel is cottage cheese.

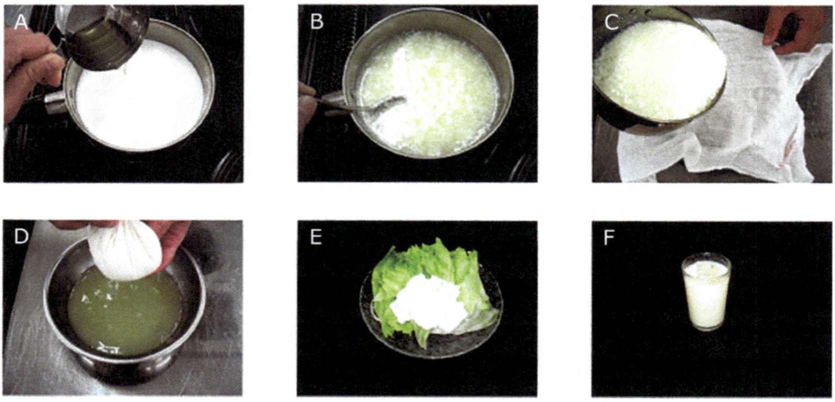

FIGURE 14.1 COTTAGE CHEESE AND WHEY. (A) POUR VINEGAR INTO HOT MILK. (B) LET IT STAND FOR 10 MINUTES. (C) STRAIN THE MILK. (D) WASH COTTAGE CHEESE A FEW TIMES AND SQUEEZE LIGHTLY. (E) PREPARATION SERVED AS PART OF A SALAD. (F) YOGURT-LIKE DRINK MADE OF WHEY, MILK, HONEY, AND LEMON JUICE.

4. Put the dish towel with the cottage cheese in water and wash it with swinging. Repeat a few times, changing the water until it is no longer turbid.
5. Squeeze the dish towel gently (Figure 14.1D) (Hint 14.6.1.2).

pH of cottage cheese = 4.81

pH of whey = 4.79

Preparation hints:

14.6.1.1 If you don't like the smell of vinegar or lemon, you can omit this process.

14.6.1.2 If you squeeze too much, the cottage cheese will become dry.

Cottage cheese and whey can be made easily at home. Cottage cheese and apples make a good salad (Figure 14.1E), while a yogurt-like drink can be made from whey, milk, honey, and lemon juice (Figure 14.1F).

14.6.2 Vanilla Ice Cream (Figure 14.2)

(Serves 5)
400 mL (1⅔ cups) milk
75 g (5 tbsp) heavy cream (45% milk fat)
18 mL (3 tbsp) skimmed milk
90 g (⅔ cup) granulated sugar
20 g (1) egg yolk
1 vanilla pod

1. Mix skimmed milk and granulated sugar in a bowl and add heavy cream and egg yolk to it, while stirring with a spatula (Figure 14.2A).
2. Remove vanilla beans from the pod and put them in a bowl; pour milk into the bowl and heat in a boiling water bath to 80°C.
3. Pour the vanilla milk (2) into the skimmed milk and sugar mixture (1), stir well with a spatula (Figure 14.2B) (Hint 14.6.2.1), and cool immediately in an ice-water bath.
4. Put the cold ice cream mix into an ice cream maker (Figure 14.2C) and follow the manufacturer's instructions (Figure 14.2D).

pH=6.44

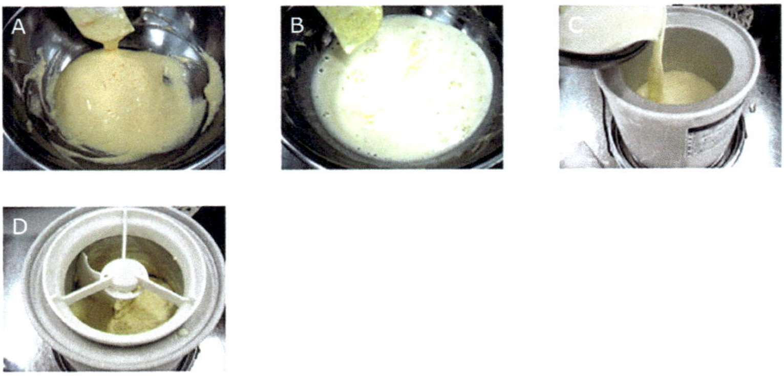

Figure 14.2 Vanilla ice cream. (A) Combine skimmed milk and granulated sugar and stir in heavy cream and egg yolk. (B) Combine vanilla milk and mix well. (C) Transfer to ice cream maker. (D) Complete preparation.

Preparation hints:

14.6.2.1 If you want to store the ice cream, heat all the ingredients in the bowl in a boiling water bath to 80°C for 10 minutes.

This recipe makes a most popular ice cream. You can add your favorite flavors or toppings.

14.6.3 Bread (Figure 14.3)

(Makes 2 loaves)

300 g (a little under 2⅕ cups) strong flour
30 g (2⅔ tbsp) casein powder
9 g (1 tbsp) white sugar
6 g (1 tsp) salt
6 g (½ tbsp) dry yeast
50 g (1) egg
25 g butter
80 mL milk
100 mL water
Strong flour, baking paper, pound-cake pan

1. Heat milk and water to around 45°C in a microwave oven; melt butter in the microwave oven; beat egg.
2. Put strong flour, casein powder, white sugar, salt, and dry yeast in a bowl (Figure 14.3A) and mix. Add the warm milk and water (1) to it and knead it (Figure 14.3B) until it no longer sticks to your hands (Hint 14.6.3.1).
3. Add the melted butter and the beaten egg (1) to the dough (2) in several steps.
4. Dust a cutting board with strong flour and knead the dough (3) well for at least 5 minutes (Figure 14.3C).
5. Put the dough (4) in a bowl, cover it with a plastic wrap (Figure 14.3D), and heat at 30°C– 35°C for 1 hour (first fermentation).
6. Poke the dough (5) with your finger. When the dent remains, the dough has properly risen (finger test) (Figure 14.3E). Divide it into

FIGURE 14.3 BREAD. (A) MIX FLOUR, CASEIN POWDER, WHITE SUGAR, SALT, AND DRY YEAST IN A BOWL. (B) COMBINE INGREDIENTS TO FORM DOUGH. (C) KNEAD THE DOUGH WELL. (D) COVER WITH PLASTIC WRAP. (E) CHECK RISING. (F) DIVIDE DOUGH INTO BALLS AND SPRINKLE WATER. (G) HEAT AND LET RISE AGAIN. (H) BAKE THE BREAD.

six to eight pieces, roll each into a ball, sprinkle water on them (Hint 14.6.3.2), and let them rest for 5 minutes (Figure 14.3F).

7. Line the pound-cake pan with baking paper, put the divided dough (6) in it, sprinkle with water, and heat to around 40°C for 30 minutes (second fermentation) (Figure 14.3G).
8. Bake the fermented dough (7) in a 190°C oven for 30–40 minutes (Figure 14.3H).

Preparation hints:

14.6.3.1 Adjust the amount of water to obtain the desired texture. Adding more water than milk makes the mixture easier to handle.

14.6.3.2 Do not let the dough become dry.

This recipe is easy to make. The amount of butter, milk, and water can be adjusted to your liking. Using a bread machine simplifies the process (Figure 14.4A–D).

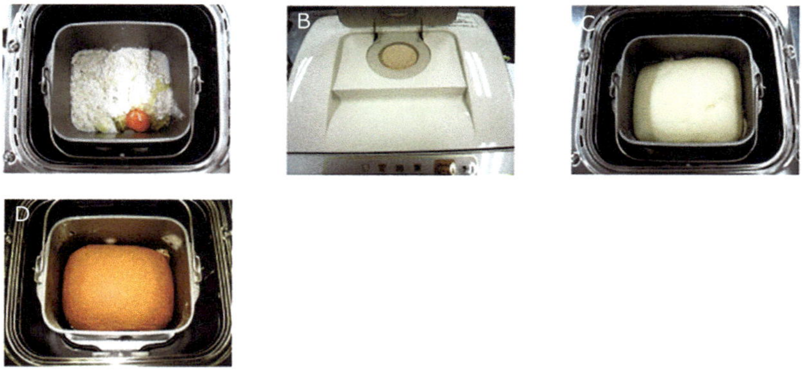

Figure 14.4 Preparation of bread using a bread machine. (A) Put all ingredients, except dry yeast, in the bread pan. (B) Place dry yeast in the dry yeast container of your bread machine. (C) Dough after second fermentation. (D) Baking.

14.6.4 Egg Pasta (Figure 14.5)

(Makes 400 g)

90 g (1 cup) whey protein concentrate (WPC) powder
170 g (1 cup) durum semolina flour
100 g (2) eggs
24 g (2 tbsp) olive oil
Salt, strong flour

1. Mix WPC, semolina flour, and a pinch of salt in a bowl.
2. Beat eggs, make a well in the flour mixture (1), pour in beaten egg and olive oil (Figure 14.5A), and mix by hand until dough becomes sticky (Figure 14.5B).
3. Dust cutting board with strong flour and knead the dough (2) until it becomes smooth (Figure 14.5C) (Hint 14.6.4.1).
4. Cover the dough (3) with a damp towel or plastic wrap and let it stand for at least 20 minutes at room temperature (Hint 14.6.4.2).
5. Cut the dough (4) into 4–5 pieces with a scraper.
6. Roll the cut dough pieces (5) out to 1 mm thickness in a pasta machine (Figure 14.5D) (Hint 14.6.4.3), cut it in 2–3 mm wide strips (Figure 14.5E), and dust with strong flour (Figure 14.5F).

FIGURE 14.5 EGG PASTA. (A) ADD BEATEN EGG AND OLIVE OIL TO A MIXTURE OF WHEY PROTEIN, SEMOLINA FLOUR, AND SALT. (B) COMBINE INGREDIENTS TO FORM THE DOUGH. (C) KNEAD. (D) ROLL THE DOUGH THROUGH A PASTA MAKER. (E) CUT 2–3 MM WIDE STRIPS. (F) DUST WITH FLOUR. (G) COOK THE PASTA IN BOILING WATER. (H) SERVE WITH YOUR FAVORITE SAUCE.

7. Boil the cut pasta (6) for 5 minutes (Figure 14.5G) and serve with your favorite sauce (Figure 14.5H).

pH = 6.75

Preparation hints:

14.6.4.1 It is easier to knead the dough using a baking machine. Knead the dough until it becomes stiffer than your earlobe.

14.6.4.2 Semolina flour granules are bigger and they take more time to absorb water.

14.6.4.3 If you do not have a pasta machine, use a rolling pin and a knife.

This recipe creates protein-rich pasta. It is a good muscle builder.

14.6.5 Cocoa Cookies (Figure 14.6)

(Makes 20–30 cookies)

30 g (a little over 4½ tbsp) milk protein concentrate (MPC) powder
120 g (a little over ⅘ cup) weak flour
10 g (1⅔ tbsp) cocoa powder
40 g (3⅓ tbsp) granulated sugar
40 g (3⅓ tbsp) vegetable oil
40–50 mL (2⅔–3⅓ tbsp) water
Weak flour

1. Mix MPC, weak flour, cocoa powder, and granulated sugar in a bowl.
2. Add vegetable oil and water to the powder mixture (1) (Figure 14.6A) and mix by hand until the dough becomes smooth (Figure 14.6B) (Hint 14.6.5.1).
3. Preheat the oven to 170°C and cover a baking tray with baking paper.
4. Dust the cutting board and rolling pin with weak flour.
5. Roll out the dough (2) to 3 mm thickness (Figure 14.6C) and cut out shapes with a cookie cutter (Figure 14.6D).

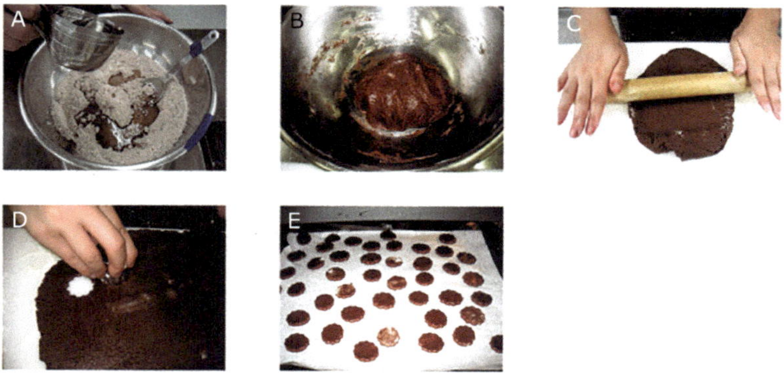

Figure 14.6 Cocoa cookies. (A) Add vegetable oil and water to the powdered mixture. (B) Mix the dough. (C) Roll out the dough. (D) Cut out cookie shapes. (E) Place the cookies on the baking tray.

6. Place the cookies (5) on the baking tray (Figure 14.6E).
7. Bake the cookies at 170°C for 17 minutes.

pH of dough=6.03

Preparation hints:

14.6.5.1 If the dough is dry and crumbly, add water.

This recipe is for hard-type cookies, which can be made in any shape. This is also a protein-rich food.

14.7 TIPS FOR THE AMATEUR COOK AND PROFESSIONAL CHEF

- Whey proteins provide economical and nutritional benefits as a substitute for eggs in cake manufacture.
- Whey proteins can replace up to 20% of the meat protein in frankfurters.
- Whey proteins can coagulate with heating. Their gels are white, turbid, and brittle.
- Disordered proteins have greater emulsifying ability than aggregated proteins.
- Liquid MPC and cream mixtures can be used for the production of reduced-fat cheeses. The use of liquid MPC for cheese-making increases throughput, enhances yield, and reduces whey volume.
- The effects of MPC incorporation in ice cream may be more apparent and more beneficial in reduced-fat or fat-free applications.

References and Further Reading

Abou El-Nour, A., Scheurer, G. J., and W. Buchheim. 1998. Use of rennet casein and milk protein concentrate in the production of spread-type processed cheese analogue. *Milchwissenschaft-Milk Sci. Int.* 53:686–90.

Alvarez, V. B., Wolters, C. L., Vodovotz, Y., and T. Ji. 2005. Physical properties of ice cream containing milk protein concentrates. *J. Dairy Sci.* 88:862–87.

Carr, A. J., Southward, C. R., and L. K. Creamer. 2003. Protein hydration and viscosity of dairy fluids. In *Advanced dairy chemistry-1: Proteins*, 3rd edn., eds. P. F. Fox and P. L. H. McSweeney, 1289–1323. New York: Kluwer Academic/Plenum Publishers.

Cayot, P., and D. Lorient. 1997. Structure-function relationships of whey proteins. In: *Food proteins and their applications*, Vol. 8, eds. S. Damodaran and A. Paraf, 225–56. New York: Marcel Dekker.

Chandan, R. 1997. *Dairy-based ingredients*. St. Paul, MN: Eagen Press.

Chiba, H., and M. Yoshikawa. 1986. Biologically functional peptides from food proteins: New opioid peptides from milk proteins. In *Protein tailoring for food and medical uses*, eds. R. E. Feeney and J. R. Whitaker, 123–53. New York: Marcel Dekker.

De Wit, J. N. 1989. The use of whey protein products. In *Developments in dairy chemistry-4*, ed. P. F. Fox, 323–46. London: Elsevier Applied Science Publishers.

Dickinson, E. 2003. Interfacial, emulsifying and foaming properties of milk proteins. In *Advanced dairy chemistry-1: Proteins*, 3rd edn., eds. P. F. Fox and P. L. H. McSweeney, 1239–60. New York: Kluwer Academic/Plenum Publishers.

Fitzgerald, R. J., and H. Meisel. 2003. Milk protein hydrolysates and bioactive peptides. In *Advanced dairy chemistry-1: Proteins*, 3rd edn., eds. P. F. Fox and P. L. H. McSweeney, 675–98. New York: Kluwer Academic/Plenum Publishers.

Fox, P. F. 1989. The milk protein system. In *Developments in dairy chemistry-4: Functional milk proteins*, ed. P. F. Fox, 1–53. London: Elsevier Applied Science Publishers.

Fox, P. F. 2003. Milk proteins: General and historical aspects. In *Advanced dairy chemistry-1: Proteins*, 3rd edn., eds. P. F. Fox and P. L. H. McSweeney, 1–48. London: Kluwer Academic/Plenum Publishers.

Francolino, S., Locci, F., Ghiglietti, R., Iezzi, R., and G. Mucchetti. 2010. Use of milk protein concentrate to standardize milk composition in Italian citric Mozzarella cheese making. *LWT—Food Sci. Technol.* 43:310–4.

Gobbetti, M., Minervini, F., and C. G. Rizzello. 2007. Bioactive peptides in dairy products. In *Handbook of food products manufacturing*, ed. Y. H. Hui, 489–517. Chichester: John Wiley & Sons.

Hilpert, H. 1984. Preparation of a milk immunoglobulin concentrate from cow's milk. In *Human milk banking*, eds. A. F. Williams and J. F. Baum, 17–28. New York: Raven Press.

Holt, C. 1997. The milk salts and their interaction with casein. In *Advanced dairy chemistry*, ed. P. F. Fox, 233–56. London: Chapman & Hall.

Hugunin, A. G. 1987. *Applications of UF whey protein: Developing new markets*, Bulletin 212, 135–44. Brussels: International Dairy Federation.

Mangino, M. E. 1992. Properties of whey protein concentrates. In *Whey and lactose processing*, ed. J. G. Zadow, 231–270. London: Elsevier Applied Science Publishers.

Mater, C., Leblanc, J. G., Martin, L., and G. Perdigon. 2003. Biologically active peptides released in fermented milk: Role and functions. In *Handbook of fermented functional foods. Functional foods and nutraceuticals series*, ed. E. R. Farnworth, 177–201. Boca Raton, FL: CRC Press.

Maubois, J.-L. 1997. *Current uses and future perspectives of MF technology in the dairy industry*, Bulletin 320, 37–40. Brussels: International Dairy Federation.

Morr, C. V. 1989. Whey proteins: Manufacture. In *Developments in dairy chemistry-4*, ed. P. F. Fox, 245–84. London: Elsevier Applied Science Publishers.

Muller, L. L. 1982. Manufacture of caseins, caseinates and co-precipitates. In *Developments in dairy chemistry-1*, ed. P. F. Fox, 315–37. London: Elsevier Applied Science Publishers.

Mulvihill, D. M. 1992. Production, functional properties and utilization of milk protein products. In *Advanced dairy chemistry*, vol. 1, ed. P. F. Fox, 369–404. London: Elsevier Applied Science Publishers.

Mulvihill, D. M., and P. F. Fox. 1994. Developments in the production of milk proteins. In *New and developing sources of food proteins*, ed. B. J. F. Hudson, 1–30. London: Chapman and Hall.

Nielsen, V. H. 1974. What exactly is whey? *Am. Dairy Rev.* 36:68–71.

Novak, A. 1996. *Applications of membrane filtration in the production of milk protein concentrates*, Bulletin 311, 26–7. Brussels: International Dairy Federation.

Nussinovitch, A. 1997. *Hydrocolloid applications. Gum technology in the food and other industries*. London: Blackie Academic & Professional.

Nussinovitch, A. 2003. *Water soluble polymer applications in foods*. Oxford: Blackwell Publishing.

Nussinovitch, A. 2010. *Plant gum exudates of the world sources, distribution, properties and applications*. Boca Raton, FL: CRC Press, Taylor & Francis Group.

Nussinovitch, A., and M. Hirashima. 2014. *Cooking innovations, using hydrocolloids for thickening, gelling and emulsification*. Boca Raton, FL: CRC Press, Taylor & Francis Group.

O'Connel, J. E., and C. Flynn. 2007. The manufacture and applications of casein derived ingredients. In *Handbook of food products manufacture*, ed. Y. H. Hui, 557–91. Chichester: John Wiley and Sons.

Otten, M. G. 1985. Whey protein concentrate: Past, present and future. In *Proceedings of IDF Symposium: New dairy products via new technology*, 107–15. Atlanta: International Dairy Federation.

Sai Manohar, R., Urmila Devi, G. R., Bhattacharya, S., and G. Venkateswara. 2011. Wheat porridge with soy protein isolate and skimmed milk powder: Rheological, pasting and sensory characteristics. *J. Food Eng.* 103:1–8.

Singh, H. 2005. Milk protein functionality in food colloids. In *Food colloids: Interaction, microstructure and processing*, ed. E. Dickinson, 179–93. Cambridge: Royal Society of Chemistry.

Southward, C. R., and N. J. Walker. 1982. Casein, caseinates and milk protein co-precipitates. In *CRC handbook of processing and utilization in agriculture*, vol. 1. *Animal products*, ed. A. Wolf, 445–552. Boca Raton, FL: CRC Press.

Swaisgood, H. E. 2003. Chemistry of the caseins. In *Advanced dairy chemistry-1: Proteins*, 3rd edn., eds. P. F. Fox and P. L. H. McSweeney, 140–201. London: Kluwer Academic/Plenum Publishers.

Teschmacher, H., and V. Brantl. 1994. Milk-protein-derived atypical opioid peptides and related compounds with opioid antagonist activity. In *β-Casomorphins and related peptides: Recent developments*, eds. V. Brantl and H. Teschemacher, 3–17. Weinheim: VCM.

Walstra, P., Geurts, T. J., Noomen, A., Jellema, A., and M. A. J. S. van Boekel. 1999. Dairy technology. In *Principles of milk properties and processes*, 3–170. New York: Marcel Dekker.

Zadow, J. G. 1986. Utilization of milk components: whey. In *Modern dairy technology*, vol. 1. *Advances in milk processing*. ed. R. K. Robinson, 273–316. London: Elsevier Applied Science Publishers.

CHAPTER 15

Other Microbial Polysaccharides
Alternan, Elsinan, and Scleroglucan

15.1 INTRODUCTION

Microbial polysaccharides are produced through well-controlled processes. Apart from the well-established microbial polysaccharides, numerous novel polysaccharides have been obtained from fungi, yeast, and bacteria as a result of the extensive research done on polysaccharide-producing strains and a better explanation of their properties and functionalities. Microbial polysaccharides have potential applications in food, and the current chapter provides relevant details related to the structure, production, functionality, and regulatory status of alternan, elsinan, and scleroglucan.

15.2 ALTERNAN

15.2.1 Manufacture and Structure

Alternan is one of two extracellular α-D-glucans that is referred to as the fraction S. It is produced by rare strains of the bacterium *Leuconostoc mesenteroides* NRRL B-1355. Synthesis of this gum is related to the enzymatic activity of alternan sucrase. This polysaccharide is produced by the hydrolysis of sucrose. The native polymer can be produced in two ways: in vivo by the bacterial culture; or in vitro from a cell-free supernatant containing alternan sucrase. The fermentation medium contains sucrose as the carbon source, a nitrogen source (i.e., liver, beef, yeast extract, or peptone), and inorganic salts. Fermentation under aerobic conditions or in a static culture is conducted for 2–3 days at temperatures of 25°C–30°C. After centrifugation and filtration (to remove bacterial cells), the gum is precipitated in ethanol, methanol, or isopropanol. Purification can be done by redissolving in water and then reprecipitating. The resultant gum is then dried to a powdery product. Alternan is a branched homopolymer, which has a unique structure containing α-(1,3) and α-(1,6)-linked D-glucose residues. Further information on its structure can be found elsewhere. This unique structure makes the polysaccharide highly soluble and less viscous. In addition, it is resistant to hydrolysis by mammalian and microbial enzymes. Its apparent average molecular mass is 10^6–10^7 Da.

15.2.2 Properties

The gum in its powdery form is white, tasteless, and not highly hygroscopic. The high solubility and low viscosity of the polysaccharide are related to its unusual structure, but the gum lacks emulsifying properties. Alternan's resistance to hydrolysis by mammalian and microbial enzymes makes it a suitable candidate for functional food. The gum solutions are stable in the temperature range of 4°C–37°C and in the pH range of 3.0–9.0.

15.2.3 Food Applications and Regulatory Status

The properties of alternan make it potentially suitable for food applications. One option is to utilize it as a non-caloric ingredient in food products that are sweetened with artificial sweeteners. Its resistance to enzymatic hydrolysis is beneficial in functional food. Another option is to use it as a fiber, bulking agent, or binder in different food items. The bacterium *L. mesenteroides* is generally regarded as safe (GRAS), and its strains are non-pathogenic and are commonly used for the preparation of many food items due to their safe history of use in the European Union.

Alternan can also be used as a thickening agent to increase the viscosity of liquids or to improve the thixotropic properties of gels. The thickening agent should be composed of alternan and at least one other thickening agent. Such compositions have a synergistic thickening effect. Alternan is particularly beneficial for thickening dietetic foodstuffs, because it is edible, but non-caloric. Foodstuffs may include sauces, meat juices, soups, dressings, dips, yoghurt, cream, full-fat milk, skimmed milk, buttermilk, sour milk, kefir, whey, mousse, jelly, pudding, spreads, jam, ice cream, baked products, and doughs. Alternan can be mixed with the dietetic foodstuff during its production or may be pre-added directly. Alternan is preferable in the form of a powder or paste (mixed with water). Alternan should be added to food in an amount that provides the desired degree of thickening. Common amounts start from 0.1% (w/w).

15.3 ELSINAN

15.3.1 Manufacture and Structure

Elsinan is an extracellular, water-soluble gum produced by *Elsinoë leucospila* and other *Elsinoë* species, such as *E. fawcetti* and *E. ampelina*. The gum is tasteless, odorless, edible, non-toxic, and biodegradable. The gum is hydrolyzed by the enzyme α-amylase.

To manufacture elsinan, the microorganisms are grown in a culture medium that contains suitable carbon (i.e., sucrose, maltose, fructose, or starch hydrolysates) and nitrogen (i.e., polypeptone, corn steep liquor, yeast extract, peptides, or amino acids) sources. Either a static or a shaken cultivation can be used. Yields of 23–25 g/L have been reported with a 5% sucrose solution. The culture should also include suitable minerals (salts), such as phosphates, potassium salts, sulfates, and magnesium. Fermentation at an initial pH of 5.0–8.0, at 20°C–30°C, for a period of 3–7 days is common. Filtration and/or centrifugation are used to remove the cells and mycelia. The gum is precipitated by methanol, ethanol, and acetone. The resultant gum is recovered by centrifugation or filtration. Further purification can be accomplished by redissolution, followed by reprecipitation, recovery, and drying. The gum is a linear polymer of α-(1,4) and α-(1,3)-linked D-glucose residues (present in a ratio of 2.6:1). Elsinan is composed of maltotriose and maltotetraose units. X-ray diffraction demonstrates a 5-fold helical structure. Its molecular mass ranges from $5–10^4$ kDa.

15.3.2 Properties

An aqueous solution of elsinan has a high viscosity. The gum is stable at a pH range of 3.0–11.0 and at temperatures of 30°C–70°C. The gum has a white hue in powdered form, and it is tasteless and colorless. The gum dissolves well in water, 0.1 N sodium hydroxide solution, 90% formic acid, formamide, or dimethyl sulfoxide. It is not soluble in methanol, ethanol, acetone, chloroform, or ethyl acetate. Elsinan forms films that are soluble in water, at or above 80°C. At temperatures <40°C, the film swells and retains its shape. Plasticizers, such as glycerin, can be added to modify the elongation of such films. Casting can also be used to prepare films that maintain their optical and mechanical properties. The flexible films are resistant to hydrophobic components such as fats and oils, and they are also glossy, transparent, colorless, edible, non-toxic, and biodegradable. Furthermore, elsinan can take on different shapes, such as granules, fiber, cloth, gauze, tube, sponge, and laminate.

15.3.3 Food Applications and Regulatory Status

Elsinan can be used as a dietary fiber to reduce serum cholesterol level under hypercholesterolemic conditions. Since elsinan can be used to create many shapes with unique properties, it can serve various purposes in the food industry. It can form films that create an oxidation barrier for perishable, oily, and variously processed food products. Currently, elsinan has no clear approval as a food additive in the United States or Europe.

15.4 SCLEROGLUCAN

15.4.1 Manufacture and Structure

Scleroglucan is an extracellular polysaccharide produced by the fungus *Sclerotium glucanicum*. Today, the term scleroglucan designates a class of glucans, having a similar structure, produced by fungi of the genus *Sclerotium*. The marketable product has a few names. In general, it is a neutral and branched homopolysaccharide of glucose. Due to its unique stability at higher temperatures, under a wide range of pH, and in the presence of different electrolytes, this gum can be used in food and other applications.

It is most commonly synthesized by *S. glucanicum* and *S. rolfsii*. The various production stages are as follows: preparation of the inoculum, controlling the composition of the growth medium, determination of the conditions under which production occurs (i.e. 28°C as optimum temperature), and controlling the formation of byproducts during the process. High yields of 60%–80% have been reported, for media enriched with a carbon source (such as glucose or sucrose). Other medium constituents include oxygen, nitrogen, phosphorus, sulfur, potassium, and magnesium. Production yields are maximized if pH is kept at 3.5 for the first stage, and then increased to 4.5 for the second stage. The gum is recovered by neutralizing the culture broth, diluting with distilled water, heating at 80°C for 30 minutes, and then homogenizing and centrifuging. Other methods of scleroglucan recovery are described elsewhere. When scleroglucan is dissolved in water it forms linear triple helices. The D-glucosidic side groups protrude and prevent aggregation. The molecular mass of the gum varies with the microbial strain, averaging $0.13–6.0 \times 10^6$ Da.

15.4.2 Properties

The properties of scleroglucan gum depend on its molecular mass, the recovery method used, and the extent of its purity. The gum is soluble in water at ambient temperatures. Upon dissolution, pure gum grades show pseudoplastic flow, and the crude products result in low-viscosity fluids. The viscosity of the gum solution is unaffected by pH in the range of 1.0–11.0, and is slightly influenced by changes in temperature. At ~7°C, thermoreversible gels are formed. In aqueous solutions, at or above pH 12.5 and at temperatures >90°C, there is a decrease in the viscosity specific rotation and the sedimentation coefficient of scleroglucan solutions. At higher electrolyte concentrations, gel formation or solution flocculation may occur. The gum is compatible with various common thickeners, such as locust bean gum and xanthan. Solutions of 1.2%–1.5% scleroglucan form sliceable, self-supporting gels at ~25°C. Very dilute solutions, at temperatures <10°C, produce diffused, structured gels that disperse upon agitation, tend to shrink, and go through syneresis after extended periods of time. Purified gum solutions at 0.1%–0.2% can suspend 5%–10% of fine powder particles. Such suspensions are pourable. Scleroglucan is also capable of producing stable oil-in-water emulsions.

15.4.3 Possible Food Applications and Regulatory Status

Scleroglucan can be used in food as a thickening, gelling, or stabilizing agent. If the cost of gum production was lower, it could have replaced xanthan in various food products, including jams, marmalades and confections, soups, dairy products, and frozen products, etc. Due to its thermal stability, it can be used in food items that are subjected to heat treatment. It can also be used in products that suffer from syneresis. In the food industry, numerous Japanese patents prove that scleroglucan can improve the quality of Japanese cakes, steamed food, rice crackers, and baked products. Scleroglucans are GRAS by the US Food and Drug Administration. Experiments with animals have not demonstrated any significant toxic effects, adverse reactions, or sensitization.

15.5 RECIPES

15.5.1 General

It is almost impossible to get samples of alternan or elsinan gums and, therefore, there are no recipes using these gums in this chapter. Moreover, the samples of scleroglucan that are available in the market are not food-grade. On one hand, numerous Japanese patents in the food industry and a few hydrocolloid texts, including famous books (see list of references), prove that scleroglucans improve the quality of frozen food, Japanese cakes, steamed food, rice crackers, and baked products. On the other hand, it is quite possible that food-grade scleroglucan will never be produced. Thus, it is misleading to suggest that scleroglucan can be currently used in food products. A few recipes which include this gum have been mentioned in this chapter, only to exemplify the claims made by some researchers about its potential to be used in food items, which will happen only when such a food-grade hydrocolloid is produced.

15.5.2 Gluten-Free Bread (Figure 15.1)

(Makes 8 cm diameter loaf)

150 g (a little over 1 cup) wheat starch (Hint 15.5.1.1)

30 g (6 tbsp) soybean flour

6 g (2 tsp) white sugar

20 g (1 tbsp) corn syrup

5 g (a little over 1 tsp) vegetable oil

1 g (½ tsp) xanthan gum powder

1 g (⅓ tsp) scleroglucan powder
11 g (a little over 1 tbsp) dry yeast
190 g (¾ cup) water
Salt (optional)

1. Heat water to around 50°C.
2. Put all ingredients (Figure 15.1A) in a bowl and mix with a spatula (Figure 15.1B).
3. Cover the dough (2) with plastic wrap and heat at 35°C for 1 hour in an oven.
4. After the fermentation (Figure 15.1C), scoop the dough with a spoon (Hint 15.5.1.2) onto the baking tray (Figure 15.1D).
5. Bake the fermented dough (4) in a 200°C oven for 20 minutes (Figure 15.1E) (Hint 15.5.1.3).

pH of dough = 5.1

Preparation hints:

15.5.1.1 Cornstarch can be used instead.

FIGURE 15.1 GLUTEN-FREE BREAD. (A) ALL INGREDIENTS. FROM TOP LEFT TO BOTTOM RIGHT: STARCH, SOYBEAN FLOUR, XANTHAN GUM POWDER, SCLEROGLUCAN POWDER, VEGETABLE OIL, DRY YEAST, SUGAR, CORN SYRUP, AND LUKEWARM WATER. (B) MIX ALL INGREDIENTS WITH A SPATULA. (C) FERMENTED DOUGH. (D) PUT THE FERMENTED DOUGH ON A BAKING TRAY. (E) BAKE AT 200°C FOR 20 MINUTES. (F) BAKED AT 210°C FOR 15 MINUTES.

15.5.1.2 The dough is very sticky, so using two spoons may work even better.

15.5.1.3 Baking at 210°C for 15 minutes leads to crunchier and browner bread (Figure 15.1F).

15.6 TIPS FOR THE AMATEUR COOK AND PROFESSIONAL CHEF

- Scleroglucan is compatible with thickeners such as xanthan.
- Scleroglucan might be useful in stabilizing oil-in-water emulsions.

REFERENCES AND FURTHER READING

Coviello, T., Palleschi, A., Grassi, M., Matricardi, P., Bocchinfuso, G., and F. Alhaique. 2005. Scleroglucan: A versatile polysaccharide for modified drug delivery. *Molecules* 10:6–33.

Giavasis, I. 2013. Production of microbial polysaccharides for use in food. In *Microbial production of food ingredients, enzymes and nutraceuticals*, eds. B. McNeil, D. Archer, I. Giavasis and L. Harvey, 413–68. Cambridge: Woodhead Publishing Limited.

Gupta, V. K., Sreenivasaprasad, S., and R. L. Mach. 2015. *Fungal bio-molecules: sources, applications and recent developments*. Hoboken, NJ: Wiley Blackwell.

Khan, T., Park, J. K., and J.-H. Kwon. 2007. Functional biopolymers produced by biochemical technology considering applications in food engineering. *Korean J. Chem. Eng.* 24:816–26.

Leathers, T. D., Hayman, G. T., and G. L. Cote. 1997. Microorganism strains that produce a high proportion of alternan to dextran. U.S. patent 5,702,942.

Misaki, A. 2004. Elsinan, an extracellular α 1,3:1,4 glucan produced by *Elsinoe leucaspila*: Production, structure, properties and potential food utilization. *Foods Food Ingredients J. Jpn.* 209:286–97.

Misaki, A., Takaya, S., Yokobayashi, K., and Y. Tsuburaya. 1980. Glucan and a process for the production thereof using Elsinoe. U.S. patent 4,202,940.

Nussinovitch, A. 1997. *Hydrocolloid applications. Gum technology in the food and other industries*. London: Blackie Academic & Professional.

Nussinovitch, A. 2003. *Water soluble polymer applications in foods*. Oxford: Blackwell Publishing.

Nussinovitch, A. 2010. *Polymer macro- and micro-gel beads: Fundamentals and applications*. New York: Springer.

Nussinovitch, A. 2010. *Plant gum exudates of the world sources, distribution, properties, and applications*. Boca Raton, FL: CRC Press, Taylor and Francis Group.

Nussinovitch, A., and M. Hirashima. 2014. *Cooking innovations using hydrocolloids for thickening, gelling, and emulsification*. Boca Raton, FL: CRC Press, Taylor and Francis Group.

Park, J. K., and T. Khan. 2009. Other microbial polysaccharides: Pullulan, scleroglucan, elsinan, levan, alternan, dextran. In *Handbook of hydrocolloids*, eds. G. O. Phillips and P. A. Williams, 592–615. Boca Raton, FL: CRC Press and Oxford: Woodhead Publishing Limited.

Pilling, J. 2013. Use of alternan as a thickening agent and thickening agent compositions containing alternan and another thickening agent. U.S. patent 8,54,041 B2.

San-Ei Chemical Industries, Ltd. 1982. Improvement of frozen food quality with sclerogum. *Japan Kokai Tokkyo Koho* 57,163,451 [Chem. Abstr. 1983, 98, 15728r].

San-Ei Chemical Industries, Ltd. 1982. Improvement of quality of Japanese cake with sclerogum. *Japan Kokai Tokkyo Koho* 57,163,442 [Chem. Abstr. 1983, 98, 15729s].

San-Ei Chemical Industries, Ltd. 1982. Improvement of steamed food quality with sclerogum. *Japan Kokai Tokkyo Koho* 163,163,441 [Chem. Abstr. 1983, 98, 15730k].

San-Ei Chemical Industries, Ltd. 1982. Rice cracker quality improvement with sclerogum. *Japan Kokai Tokkyo Koho* 57,163,440 [Chem. Abstr. 1983, 98, 15731m].

San-Ei Chemical Industries, Ltd. 1982. Quality improvement of bakery products with sclerogum. *Japan Kokai Tokkyo Koho* 57,163,432 [Chem. Abstr. 1983, 98, 15732n].

Survase, S. A., Saudagar, P. S., Bajaj, I. B., and R. S. Singhal. 2007. Scleroglucan: Fermentative production, down stream processing and applications. *Food Technol. Biotechnol.* 45:107–18.

Wang, Y., and B. McNeil. 1996. Scleroglucan. *Crit. Rev. Biotechnol.* 16:185–215.

Zhu, D., Damodaran, S., and J. A. Lucey. 2010. Physicochemical and emulsifying properties of whey protein isolate (WPI)–dextran conjugates produced in aqueous solution. *J. Agric. Food Chem.* 58:2988–94.

CHAPTER 16

Pullulan

16.1 INTRODUCTION

Pullulan, a glucan polymer, is used for the development of copolymers, permeable barrier coatings, nanofibers, nanospheres, nanogels, and biofilms. The potential uses of these products range from protective food-wrapping materials to biocompatible tissue engineering and drug-delivery materials and structures. In 1938, early observations on pullulan formation by *Aureobasidium pullulans* were described. In 1958, isolation and characterization of the polymer began. During the 1960s, the basic structure of pullulan was resolved. In 1961, the enzyme pullulanase—which specifically hydrolyzes the (1,6) linkages in pullulan and converts the polysaccharide almost quantitatively to maltotriose—was discovered. Hayashibara Company Ltd., in Okayama, Japan began the commercial production of pullulan in 1976.

16.2 SOURCES AND MANUFACTURE

Pullulan is an extracellular polysaccharide. It can be produced via fermentation by *A. pullulans,* which is a ubiquitous yeast-like fungus that can be found in different environments (e.g., soil, water, air, and limestone). *A. pullulans* is highly important in biotechnology for the production of different enzymes, siderophores, and pullulan. The latter is produced on a medium containing 10%–15% starch hydrolysates and basal salts. The medium, which also contains peptone and phosphate, is aerated and the pH is adjusted to 6.5 and the temperature is kept at 30°C. During fermentation, the pH decreases to ~3.5 and maximal culture growth occurs within 75 hours, while optimum yields of pullulan are obtained within ~100 hours. The polysaccharide is marketed as an odorless, flavorless, stable, white powder. Commercial samples contain more than 90% hydrocolloids. Impurities include mono-, di-, and oligosaccharides.

Upon completion of the fermentation, microfiltration is employed to remove the fungal cells and then the filtrate is heat sterilized. Further removal of melanin and other impurities is achieved by using activated charcoal. The decolorized filtrate is cooled, deionized, and then concentrated to a solid content of ~12%. Further repeated treatment is performed with activated carbon, and diatomaceous earth is used for further filtration. This is followed by concentration to ~30% solid content and then a drum dryer is used to dry the concentrated filtrate. The pullulan, in its dried form, is further pulverized to obtain the required particle size. The powder is packaged in sterilized polyethylene bags. Yields >70% have been reported. Temperature, pH, and type of substrate and strain influence the manufacturing process.

16.3 STRUCTURE AND PROPERTIES

Pullulan is a linear homopolysaccharide, which consists of maltotriose and maltotetraose with α-(1,6) and α-(1,4) linkages in regular alternation. Sometimes, depending upon culture conditions and cultivar differences, a minute proportion of α-(1,3) linkages is also present. The molecular mass of the gum ranges from 10–3,000 kDa and these values depend upon manufacturing conditions. Further information on pullulan structure and its resemblance to starch amylopectin and maltodextrin can be located elsewhere.

The unique linkage pattern in the pullulan molecule confers structural flexibility and enhanced solubility. At low concentrations, its solutions have high viscosity. Properties such as its ability to be molded into fibers and to form strong oxygen-impermeable films are related to this characteristic. The gum is neither toxic nor mutagenic. In addition, it is odorless and tasteless and produces a whitish powder. The gum is edible, soluble in cold and hot water, and insoluble in organic solvents, except in dimethylformamide and dimethylsulfoxide. Aqueous solutions of pullulan are viscous and clear. The gum does not form gels. The viscosity of pullulan solution is lower than that of guar gum solution. Its viscosity is not greatly affected by heating, pH changes, or the presence of most metal ions, including sodium chloride.

16.4 FOOD APPLICATIONS AND REGULATORY STATUS

Pullulan has special uses in food. It is regarded as a dietary fiber. In addition, it promotes the growth of beneficial bifidobacteria. Pullulan can be used as a substitute for starch in pastas and baked goods. Other possible uses are as fillers in beverages and sauces.

Due to its ability to produce water-soluble, fat-resistant films, having limited permeability to both moisture and oxygen, pullulan can be used as a replacement for gelatin in the production of capsule shells for dietary supplements. Pullulan is an excellent film-former, producing a film that is colorless, tasteless, odorless, transparent, flexible, highly impermeable to oil, and heat-sealable with good oxygen-barrier properties. The use of pullulan in film-forming and coating materials, by itself or in blends with different polysaccharides, has long been reported. Its effect on the properties of protein-based edible films has been intensively studied with soy protein isolate (Chapter 18) and caseinate (Chapter 14). The structural and physical features of films obtained in the presence or absence of pullulan were studied, and such microstructure analysis demonstrated good compatibility of glycerol (the plasticizer) with whey protein isolate (Chapter 14). Also, the absence of empty spaces in the whey protein isolate–pullulan film indicated good compatibility between the two components. The use of a whey protein isolate–pullulan coating of roasted chestnuts, in combination with freeze-drying and low-temperature storage, was effective in controlling the overgrowth of spoilage organisms and surface discoloration. The moisture content of roasted chestnuts appears to be the major factor that limits their storage life. Thus, there is a need to improve the moisture-barrier properties through the incorporation of lipids, such as sunflower oil or fatty acids, in the coating formulation. Thus, the whey protein isolate–pullulan coating combined with freeze-drying and low-temperature storage can be an alternative strategy to minimize the significant losses of harvested chestnuts. Analysis by the scanning electron microscopy showed microstructures that indicated good compatibility between the whey protein isolate–pullulan coating and the fruit surface, i.e., the integrity was perfect. Pullulan can also be used for the production of edible, flavored films, jams and jellies, confectionery, and different fruit and meat products. Pullulan, gelatin, lactose, and their mixtures were added to whole egg prior to drying; results showed that storage and the addition of carbohydrate and/or protein-based materials affect the functional properties of egg powder. The addition of pullulan gave rise to high-expanding and permanent foam; however, its gelling properties (gel firmness and water-holding capacity) showed lower values than that of the other egg powders.

Pullulan is used as a texturizing agent in chewing gum and as a foaming agent in milk-based desserts. Pullulan is edible and can be consumed as part of the food that it coats. Such coatings reduce oxidation of the coated commodity and protect it against mold. In Japan, pullulan is used in baked goods and beverages. It is used as a binder for seasoning, as a sheet for wrapping various food items, and as a packaging material for instant noodles. Pullulan is also used as a texturizer for *tofu* and meat products and as an additive in low-calorie foods and beverages. The United States Food and Drug Administration (FDA) categorizes pullulan as a GRAS additive for optional use in food, particularly in meat products.

16.5 RECIPES

16.5.1 Almond Cookies (Figure 16.1)

(Makes 10–15 cookies)

A { 10 g (2 2/3 tsp) pullulan powder (Hint 16.5.1.1)
  90 mL (a little over 1/3 cup) watver

10–30 shelled whole almonds
150 g (a little over 1 cup) all-purpose flour
10 g (5/6 tbsp) baking powder
0.7 g (a little under 1/5 tsp) pullulan powder (Hint 16.5.1.2)
60 g butter
60 g (a little over 1/3 cup) white sugar

Figure 16.1 Almond cookies. (A) Prepare pullulan solution and soak almonds in it. (B) Mix softened butter, and white sugar. (C) Add egg and vanilla flavoring. (D) Combine the ingredients into a mixture. (E) Shape into balls. (F) Put one or two almonds on each cookie. (G) Bake.

1 egg
Vanilla flavoring
Water

1. Make the pullulan solution using ingredients in A. Put 10 g pullulan powder and water in a pan. Heat while stirring with a whisk, until the pullulan dissolves (at 90°C), and then soak almonds in it (Figure 16.1A).
2. Mix all-purpose flour, baking powder, and 0.7 g pullulan powder and sieve.
3. Soften butter at room temperature or in a microwave oven. Then mix with white sugar in a bowl (Figure 16.1B) (Hint 16.5.1.3), add egg and vanilla flavoring (Figure 16.1C), and mix well.
4. Add the powder mixture (2) to the butter mixture (3) and mix well (Figure 16.1D); make balls of 3 cm diameter (Figure 16.1E) (Hint 16.5.1.4).
5. Drain the almonds (1) and put one or two on each cookie ball (4) (Figure 16.1F).
6. Preheat oven to 200°C and bake cookies for 8 minutes (Figure 16.1G).

pH of pullulan solution = 8.25

pH of cookie batter = 6.07

Preparation hints:

16.5.1.1 This pullulan is used as a binding agent, so that the almonds stick to the cookies.

16.5.1.2 This pullulan is used as a water-retention agent.

16.5.1.3 Mix well until it turns white.

16.5.1.4 You can also make a 3 cm diameter cylinder to form cookies; refrigerate cylinder until firm and then cut 5 mm thick slices with a knife and bake.

This is a basic recipe for simple cookies; you can add chocolate, sesame seeds, dried fruit, or your favorite ingredients. Substituting lard for butter will produce Chinese cookies.

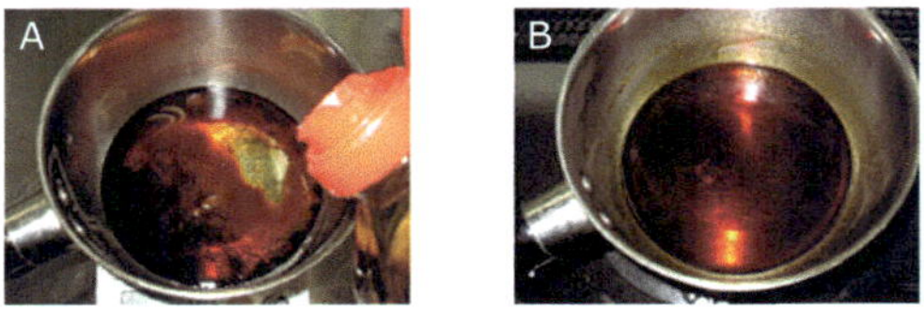

FIGURE 16.2 *TERIYAKI* SAUCE. (A) MIX PULLULAN AND WHITE SUGAR AND ADD SOY SAUCE AND *MIRIN*. (B) HEAT THE MIXTURE UNTIL THE POWDER DISSOLVES.

16.5.2 *TERIYAKI* SAUCE (FIGURE 16.2)

(Makes 120 g)
2 g (a little over ½ tsp) pullulan powder (Hint 16.5.2.1)
27 g (3 tbsp) white sugar
72 mL (4 tbsp) soy sauce
45 mL (2½ tbsp) *mirin*

1. Mix pullulan and white sugar in a pan and then add soy sauce and *mirin* (Figure 16.2A).
2. Heat the mixture (1) until the powder dissolves (65°C–70°C) (Figure 16.2B).

pH = 4.72

Preparation hints:
16.5.2.1 This pullulan is used as an adhesive and gloss agent.

Teriyaki is a traditional Japanese sauce that can be used for *teriyaki* chicken, pork, fish, and the likes. Given below is a recipe for *teriyaki* chicken.

16.5.3 *TERIYAKI* CHICKEN (FIGURE 16.3)

(Serves 5)
300 g chicken*

* You can use any chicken parts.

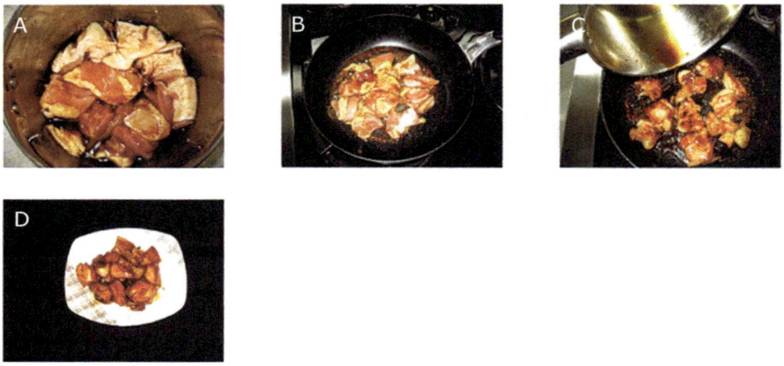

Figure 16.3 *Teriyaki* chicken. (A) Marinate cut chicken in *teriyaki* sauce. (B) Heat vegetable oil and fry the marinated chicken on high heat. (C) Turn the pieces over and cover with remaining sauce. (D) Serve.

100 mL *teriyaki* sauce
Vegetable oil

1. Cut chicken into bite-sized pieces and marinate in *teriyaki* sauce for 20 minutes (Figure 16.3A).
2. Heat vegetable oil in a frying pan and fry the marinated chicken (1) on high heat (Figure 16.3B); turn the chicken pieces over, cover with the remaining sauce (Figure 16.3C), and fry until well done (Figure 16.3D).

16.6 TIPS FOR THE AMATEUR COOK AND PROFESSIONAL CHEF

- Pullulan solution is a strong adhesive, which can glue dry food.
- Addition of pullulan inhibits drip for frozen foods.

References and Further Reading

Gounga, M. E., Xu, S.-Y., and Z. Wang. 2007. Whey protein isolate-based edible films as affected by protein concentration, glycerol ratio and pullulan addition in film formation. *J. Food Eng.* 83:521–30.

Gounga, M. E., Xu, S.-Y., Wang, Z., and W. G. Yang. 2008. Effect of whey protein isolate–pullulan edible coatings on the quality and shelf life of freshly roasted and freeze-dried Chinese chestnut. *J. Food Sci.* 73:E155–61.

Joint FAO/WHO Expert Committee on Food Additives (JECFA). 2005. Pullulan, chemical and technical assessment (CTA), WHO Food Additives Series, World Health Organization, 65th Meeting, Geneva.

Khan, T., Park, J. K., and J.-H. Kwon. 2007. Functional biopolymers produced by biochemical technology considering applications in food engineering. *Korean J. Chem. Eng.* 24:816–26.

Koc, M., Koc, B., Susyal, G., Yilmazer, M. S., Bagdatlioglu, N., and F. Kaymak-Ertekin. 2011. Improving functionality of whole egg powder by the addition of gelatin, lactose, and pullulan. *J. Food Sci.* 76:S508–15.

Leathers, T. D. 2003. Biotechnological production and applications of pullulan. *Appl. Microbiol. Biotechnol.* 62:468–73.

Nussinovitch, A. 1997. *Hydrocolloid applications. Gum technology in the food and other industries.* London: Blackie Academic & Professional.

Nussinovitch, A. 2003. *Water soluble polymer applications in foods.* Oxford: Blackwell Publishing.

Nussinovitch, A. 2010. *Plant gum exudates of the world sources, distribution, properties and applications.* Boca Raton, FL: CRC Press, Taylor & Francis Group.

Nussinovitch, A., and M. Hirashima. 2014. *Cooking innovations, using hydrocolloids for thickening, gelling and emulsification.* Boca Raton, FL: CRC Press, Taylor & Francis Group.

Park, J. K., and T. Khan. 2009. Other microbial polysaccharides: Pullulan, scleroglucan, elsinan, levan, alternan, dextran. In *Handbook of hydrocolloids*, eds. G. O. Phillips and P. A. Williams, 592–615. Boca Raton, FL: CRC and Oxford: Woodhead Pulblishing Limited.

CHAPTER 17

Soluble Soybean Polysaccharide

17.1 INTRODUCTION

Soluble soybean polysaccharide (SSPS) is a water-soluble polysaccharide extracted from soybean. It consists mostly of the dietary fiber of the soybean cotyledon (i.e., a significant part of the embryo within the seed). The polysaccharide has been sold under the brand name "Soyafiber-S" since 1993. "*Okara*" is an insoluble residue obtained after oil and protein extraction. It is obtained during the manufacture of soy protein isolate, soymilk, and *tofu*. *Okara*, which contains ~80% moisture, is further treated to extract the SSPS.

17.2 MANUFACTURE AND STRUCTURE

SSPS is extracted from the *okara* by heating it under weakly acidic conditions. The extraction technique produces a variety of SSPSs, which differ in their physical properties and functional abilities.

Following the extraction, refining, pasteurization, and spray-drying are performed to obtain the final product. The processing does not involve the use of any toxic reagents. The most common type of SSPS (i.e. Soyafiber-S-DA100) is composed of ~66.2% dietary fiber, 9.2% crude protein, and 8.6% crude ash. The main sugar constituents are galactose, arabinose, and galacturonic acid; nevertheless, various other sugars, such as rhamnose, fucose, xylose, and glucose are also present. Various SSPSs are marketed according to their functionality. Molecular mass distribution of SSPS was determined by the method of gel permeation chromatography using refractive index and multiangle laser light scattering detection. Three components of 550, 25, and 5 kDa were detected. The mean value is estimated to be several hundreds of kilodaltons. The major component of SSPS has three types of units in the main backbone, which consists of homogalacturonan and rhamnogalacturonan. Further information on the structure can be located elsewhere. SSPS has a compact globular structure and its radius of gyration is ~23.5 ± 2.8 nm.

17.3 CHARACTERISTICS

SSPS can be metabolized only to a limited extent. It is modified by enteric bacteria to an organic acid. This polysaccharide has the ability to shorten gastrointestinal transit time effectively. It is not able to produce gels. In comparison with other hydrocolloids, it has a low viscosity, which favors concentration values >30%. Another important characteristic is that its viscosity is affected only by heating or by the inclusion of acids or salts to a limited degree. The adhesive strength of SSPS films was compared to that of the films produced from pullulan and gum arabic. Adhesive strengths of 46.6, 40.5, and 30.7 kg force/cm^2 were recorded, respectively. In other words, SSPS films have the highest adhesive strength. Therefore, it was concluded that SSPS can be utilized as a binder in snacks and cereals (i.e., dried food), as well as in paper, wood, or glass applications. The high tensile strength of SSPS (~540 kg force/cm^2) makes it an ideal candidate for manufacturing films that can be used for different applications.

SSPS might prevent the oxidation of oils. The antioxidant effect of SSPS-S-LA200 on soybean oil, stored at 60°C, was evaluated. A considerable and promising decrease in peroxide values from ~60 meq/kg (for 5% added gum arabic)–~25 meq/kg (for 5% added SSPS) was detected.

17.4 FUNCTIONAL PROPERTIES

SSPS has many functions and applications. Nevertheless, it is important to note that each type of Soyafiber-S has its own functions. The polysaccharide type DN was developed as a stabilizer for protein particles, under acidic conditions. The DN type can be used as a soluble dietary fiber in fortified food, as a stabilizer in potable yogurt, ice cream, acidic desserts, and sour cream, as an adhesive agent, for film forming, and in baked goods, meat products, cream sauce, and in *kamaboko* for its softening effect. Type DA100 SSPS gives improved flavor as compared to DN. Type LN was developed as an emulsifier. Type LA200 gives improved flavor as compared to LN, and it can be used for powdering bases. Type EN100 has improved stability as compared to LN in suspension; it was developed for flavor emulsions. Type RA100 has an excellent color and a high viscosity. It can be used for cooked rice, edible films, and coatings. Finally, type HR was developed as a low-viscosity emulsifier.

17.5 FOOD APPLICATIONS AND REGULATORY STATUS

SSPSs can stabilize protein particles under low pH conditions without increasing viscosity. Acidified milk drinks comprise a large variety of manufactured goods, from those frequently produced from fermented milk with stabilizers and sugar to those prepared by direct acidification with fruit juices and/or acids. The pH of these products ranges from 3.4–4.6, and due to the instability of caseins (Chapter 14) in this range, a stabilizer is required to prevent protein aggregation and to achieve optimal mouthfeel. Stabilizers are important for controlling properties such as texture, viscosity, and mouthfeel; correct ingredient formulation and appropriate processing eventually influence consumer acceptability. When SSPS is used in acidic milk drinks, a "light" taste can be achieved with no sticky mouthfeel and a low viscosity is maintained. In products with pH < 4.0, the stabilizing ability of SSPS is at its best. In principle, this ability can be beneficial in desserts and ice creams. SSPS has low reactivity with calcium ions and, therefore, it is highly beneficial in milk products, even if applied during an initial stage of the processing. The stabilization by SSPS has been hypothesized to occur through the mechanism described here. The galacturonic acid in the polysaccharide might be located within the main backbone of the molecule. Anionic groups in the backbone bind to the surface of cationic protein particles and the hydrophilic layers of the polysaccharide, which coat the protein particles, eliminate aggregation by steric repulsion. Both SSPS and high-methoxy pectin (HMP) can stabilize casein (Chapter 14) particles in low-pH milk. SSPS functionality seems to complement that of pectin, because at pH 4.6, pectin stabilizes acid dispersions, whereas under acidic conditions (pH 3.2–4.0), SSPS sorbs to the surface of casein micelles and prevents aggregation. The differences in functionality of the two polysaccharides result from the dissimilar molecular structures. The interactions of HMP and SSPS with sodium caseinate-stabilized emulsions were investigated. At pH 6.8, both polysaccharides were negatively charged and did not sorb to the caseinate-coated droplets, due to electrostatic repulsion. While destabilization occurred at low polysaccharide concentrations, probably via bridging flocculation, acid-induced aggregation of the oil droplet was completely prevented by 0.2% SSPS or HMP. However, the interactive behavior of SSPS during acidification was different from that of HMP.

SSPS is able to emulsify flavors. Such emulsions can contain, as their main ingredients, an oil phase (e.g., lemon oil), glycerol, citric acid, and a stabilizer, such as gum arabic or SSPS. Flavor emulsions are prepared by the method described here. First, the polysaccharide and glycerol are dissolved and mixed in water. The pH of the solution is adjusted to ~4.0 by addition of citric acid. Then the oil phase is poured into the solution and mixed at 8000 rpm for 30 minutes at 35°C. The fluid is homogenized twice at 35°C at 150 kg/cm^2. The inclusion of SSPS was observed to be beneficial for obtaining good suspension stability, when added in an amount lower than gum arabic. It is possible that the protein fraction that associates with SSPS is responsible for anchoring the carbohydrate moieties of the polysaccharide onto the oil–water interface. SSPS stabilizes oil droplets within the emulsion, presumably through steric repulsion. The hydrophilic portion of the polysaccharide forms a hydrated layer of ~30 nm, which prevents coalescence of droplets.

Emulsions prepared with SSPS are stable in a pH range of 3.0–7.0. The emulsifying properties of SSPS are not affected by pH or ionic strength. When compared to literature data on the emulsifying properties of gum arabic or modified starch, SSPS is found to present similar behavior in acidic, concentrated, and dilute emulsions. However, lower amounts of SSPS are needed to stabilize acidic emulsions as compared to those reported for gum arabic and modified starch. Stable oil-in-water emulsions were produced in which β-carotene was incorporated in oil droplets that were surrounded by multilayer interfacial membranes. Emulsions were prepared using a two-stage process by the means of homogenization, which relied on the adsorption of chitosan (Chapter 5) to the anionic droplets coated with SSPS. The viscoelastic behavior of the preparation was enhanced by the adsorption of chitosan onto the SSPS-coated droplet surfaces. β-carotene emulsions were prepared using a two-stage homogenization process and adsorption of chitosan to the anionic droplets coated with SSPS. The molecular weight of chitosan had a significant impact on the heat and light stability of β-carotene in emulsions. SSPS-stabilized emulsion was more stable when there was adsorption of medium-molecular-weight chitosan, as compared to low- and high-molecular-weight chitosan. Canthaxanthin was microencapsulated by spray-drying using SSPS as a wall material. SSPS showed very good microencapsulation ability owing to its emulsification properties. The best ratio of core-to-wall was 0.25:1 because the microcapsules prepared with this ratio had the smallest sized droplets and microcapsules, and they also showed the highest microencapsulation efficiency and the lowest loss during processing.

SSPS has the ability to stabilize foams. This ability was proven by adding 0.2% polysaccharide to a 2% aqueous solution of hydrolyzed soy protein. Another very interesting effect of SSPS is its ability to serve as an antisticking agent for cooked rice and noodles. Since rice boiled with SSPS has a hard texture, additional water can be included in the preparation, and the yield of the cooked rice is therefore increased. A similar principle could be applied for cooked noodles. SSPS maintains the texture of cooked noodles for many hours. To achieve such an effect, the boiled noodles should be dipped in or sprayed with an aqueous solution of SSPS to prevent sticking. Another option is to cook the noodles in the aqueous SSPS solution. In the above cases, the addition of SSPS eliminates the need to add oil to prevent stickiness in cooked noodles, spaghetti, or chow mein. It is assumed that SSPS coats the surface of the cooked product and the coating layer prevents stickiness.

Increasing the consumption of dietary fiber has many important health benefits: for example, reduction in blood cholesterol, reduced risk of diabetes, and improved laxation. SSPS is a dietary fiber extracted and refined from *okara*, a byproduct of soy manufacturing (Chapter 18). It has been added to three types of dairy-based products—thickened milkshake-style beverages, puddings, and low-fat ice cream—in the maximum amount possible, without over-texturing the food. Dairy beverages fortified with 4% SSPS and 4% SSPS-fortified puddings come in the range of commercial products. From sensory analyses, 4% SSPS-fortified dairy beverage with 0.015% κ-carrageenan, 4% SSPS-fortified pudding with 0.1% κ-carrageenan, and 2% SSPS-fortified low-fat ice cream gained the highest scores in consumer hedonic ratings.

SSPS was also included as a dietary fiber in various food items, as a softener in baked goods, as a stabilizer of inorganic particles under acidic conditions, and for many other industrial applications. In Japan, SSPS is regarded as a food ingredient and a food additive, with no limitation to its application. The additive must be labeled as a "soybean polysaccharide" or a "soybean hemicellulose" in accordance with supplement 3 of the Food Sanitation Law of 1996. The additive can also be labeled as "soybean fiber", if it is used as such in the food items. In the United States, SSPS is categorized as self-affirmed generally regarded as safe (GRAS).

17.6 RECIPES

SSPS is not used in the kitchen at present, but it is widely used in the food industry, especially in lactic-fermented drinks (Figure 17.1A) and instant noodles (Figure 17.1B).

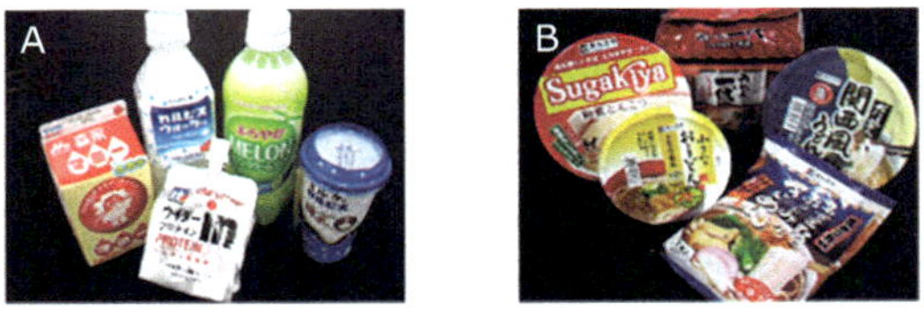

FIGURE 17.1 THE INDUSTRIAL INCLUSION OF SSPS IN LACTIC-FERMENTED DRINKS (A) AND INSTANT NOODLES (B).

17.6.1 YOGURT DRINK (FIGURE 17.2)

(Serves 5)

340 g yogurt
3 g (a little over 1 tsp) SSPS powder
20 g (4 tbsp) skimmed milk
60 g (5 tbsp) granulated sugar
450 mL (1⅘ cups) water
50 mL (3⅓ tbsp) lemon juice

1. Mix SSPS, skimmed milk, and granulated sugar in a pan, stir in water with a whisk (Figure 17.2A), heat to boiling, and cool (Figure 17.2B).
2. Blend in yogurt and lemon juice with a hand mixer for 3 minutes (Figure 17.2C) (Hint 17.6.1.1).
3. Refrigerate the yogurt drink (2) until about 80% of the bubbles disappear.
4. Mix the yogurt drink (3) and pour it into glasses (Figure 17.2D).

pH = 3.80

Preparation hints:

17.6.1.1 The yogurt drink contains many bubbles, and its volume increases during stirring so a larger pan or bowl should be used to avoid overflow.

This recipe is best served cold because it is sour. The pH needs to be lowered so that the SSPS can stabilize in the milk.

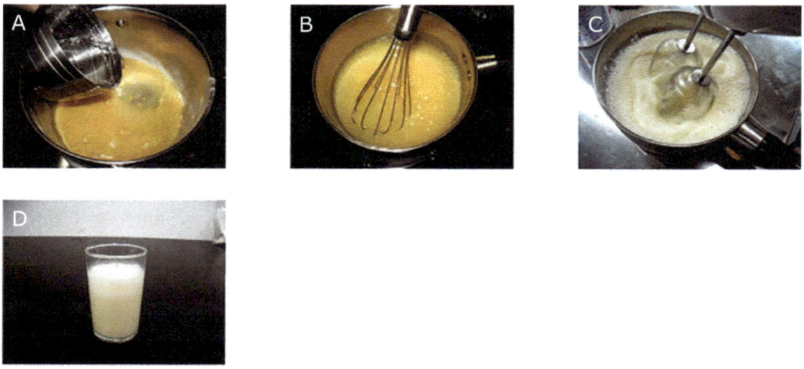

FIGURE 17.2 YOGURT DRINK. (A) ADD WATER TO A MIXTURE OF SSPS, SKIMMED MILK, AND GRANULATED SUGAR WHILE STIRRING. (B) BRING TO A BOIL AND THEN COOL. (C) ADD YOGURT AND LEMON JUICE TO SSPS SOLUTION AND MIX. (D) SERVE.

17.6.2 BOILED PASTA (FIGURE 17.3)

(Serves 5)

7.5 g (a little under 1 tbsp) SSPS powder (1% of water)
750 mL (3 cups) water
400 g pasta*
4 L water
20 g (a little over 1 tbsp) salt

1. Mix SSPS and 750 mL water and heat to 60°C while stirring with a whisk (Figure 17.3A).
2. Heat 4 L water with salt and bring to a boil. Add pasta to the boiling water (Hint 17.6.2.1) and boil until done (Figure 17.3B).
3. Remove pasta from the heat and soak in the SSPS solution (1) for 1 minute (Figure 17.3C).

pH of SSPS solution = 6.74

* You can use any type of pasta.

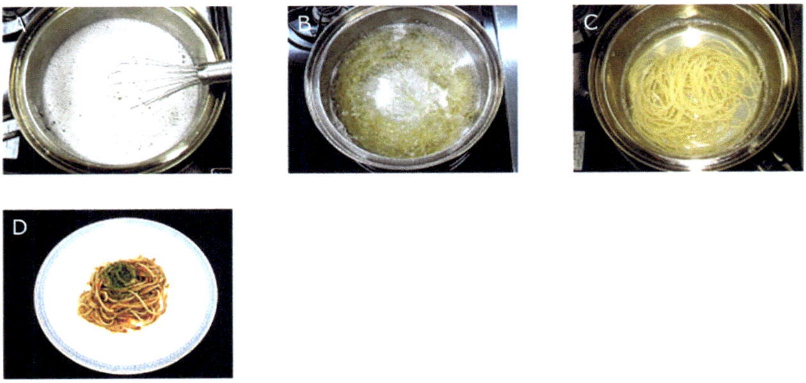

FIGURE 17.3 BOILED PASTA. (A) HEAT A MIXTURE OF SSPS AND WATER TO 60°C WHILE STIRRING. (B) BOIL SALTED WATER AND ADD PASTA. (C) SOAK THE PASTA IN THE SSPS SOLUTION. (D) SERVING SUGGESTION FOR PASTA.

Preparation hints:

17.6.2.1 For successful pasta preparation, use lots of water, add salt to the water, and add the pasta to the boiling water. This way the pasta will not stick together.

This recipe is used for retort-packed or frozen food. It is good for pasta with sauce (Figure 17.3D) or *gratin*.

17.6.3 STEAMED RICE (FIGURE 17.4)

(Serves 5)

400 g Japonica rice*

4 g (1½ tsp) SSPS powder

600 mL (2⅖ cups) water

1. Wash rice (Figure 17.4A) (Hint 17.6.3.1).
2. Put the washed rice (1) and SSPS in a bowl and add water so that the total weight of rice, SSPS, and water is 1000 g (Hint 17.6.3.2). Soak for 30–60 minutes (Figure 17.4B).

---

* Non-glutinous rice.

FIGURE 17.4 STEAMED RICE. (A) RINSE RICE. (B) PLACE THE RICE, SSPS, AND WATER IN A PAN. (C) SERVE.

3. Cook the rice following the rice cooker instructions (Hint 17.6.3.3). Serve the steamed rice (Figure 17.4C).

pH of SSPS solution = 6.97

Preparation hints:

17.6.3.1 The purpose of washing the rice is to remove rice bran and dust. Pour lots of water on the rice in a bowl. Then mix and rub the rice, pour out dirty water, and repeat this process 3 to 4 times.

17.6.3.2 While it is being washed, the rice absorbs an amount of water that is equivalent to 10% of its weight. Therefore, the total weight should be measured.

17.6.3.3 If you do not have a rice cooker, you can make the steamed rice in a pan on a stove as follows.

How to make steamed rice without using a rice cooker (Figure 17.5)

1. Heat rice and water at high heat until water boils (about 8 minutes) (Figure 17.5A).
2. Reduce the heat to medium so that the water continues to boil until it is absorbed (Figure 17.5B) (about 5 minutes).
3. Reduce the heat to low until the water completely evaporates (about 15 minutes) (Figure 17.5C).
4. Remove from heat and steam without opening the lid for 10 minutes (Figure 17.5D).
5. Open the lid and mix the rice with a spatula (Figure 17.5E).

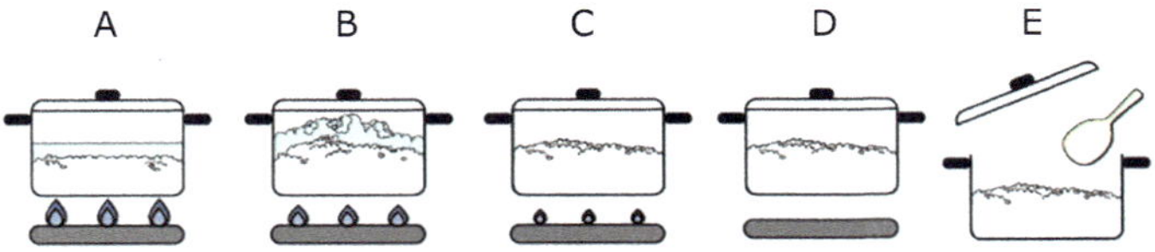

FIGURE 17.5 Preparation of steamed rice without a rice cooker. (A) Heat rice and water on high heat until it boils. (B) Reduce the heat to medium and continue boiling until water is absorbed. (C) Reduce heat to low until the water completely evaporates. (D) Remove from heat and steam without opening the lid. (E) Open the lid and mix the rice.

This recipe is not usually used at home, but it is used in the food industry to make non-sticky rice. It is good for making butter rice or fried rice. The recipe for fried rice is given below.

17.6.4 Fried Rice (Figure 17.6)

(Serves 5)

880 g steamed rice (Hint 17.6.4.1)
100 g sliced pork*
50 g leek or long onion†
40 g (3⅓ tbsp) lard or sesame oil

A:
- 2 eggs
- 1 g (1/5 tsp) salt
- 8 g (2/3 Tbsp) vegetable oil

5 g (a little under 1 tsp) salt
White pepper, soy sauce
Eggs, vegetable oil

* You can also use other meat, such as beef, chicken, sausage, or seafood such as shrimp, clams, or squid.

† You can use other vegetables, such as onion, carrot, green peas, corn, etc.

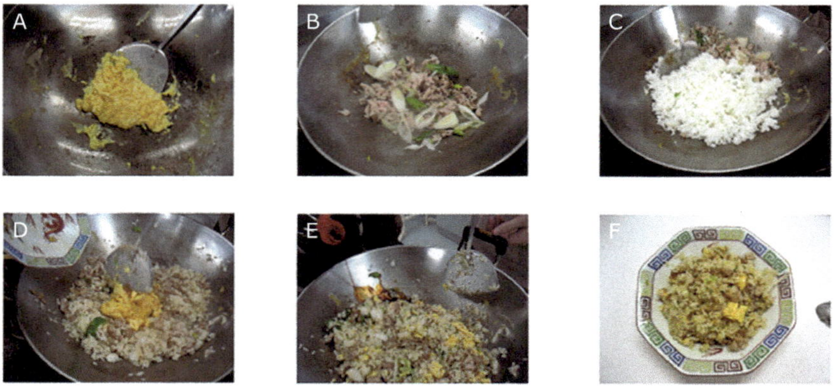

FIGURE 17.6 FRIED RICE. (A) MAKE SCRAMBLED EGGS. (B) HEAT LARD IN FRYING PAN, ADD CUT PORK AND LEEK, AND THEN STIR-FRY. (C) ADD STEAMED RICE AND LARD. (D) ADD THE SCRAMBLED EGG TO THE STIR-FRY. (E) POUR SOY SAUCE AROUND THE EDGE OF THE FRYING PAN. (F) SERVE.

1. Cut the pork into bite-sized pieces and the leek into 1 cm pieces.
2. Make scrambled eggs using ingredients in A. Beat eggs and add salt; heat vegetable oil in a frying pan, fry the beaten eggs (Figure 17.6A), and remove to a dish.
3. Heat half of lard in a frying pan and add the cut pork (1); stir-fry it until it is done, add the cut leek (1), and stir-fry (Figure 17.6B).
4. Put steamed rice and remaining lard in the frying pan, season with salt and pepper (Figure 17.6C), stir-fry, and add the scrambled eggs (2) (Figure 17.6D) (Hint 17.6.4.2).
5. Pour soy sauce around the edge of the frying pan (Figure 17.6E), mix it in, remove from heat, and serve (Figure 17.6F).

Preparation hints:

17.6.4.1 400 g of rice gives about 880 g steamed rice (about 2.2–2.3 times the original weight).

17.6.4.2 Use high heat and stir-fry quickly so that the steamed rice does not become sticky.

17.7 TIPS FOR THE AMATEUR COOK AND PROFESSIONAL CHEF

- SSPS can be dissolved in water at over 60°C.
- SSPS prevents the aggregation of milk casein under acidic conditions (pH 3.2–4.0).
- Emulsions with SSPS are stable in the pH range of 3.0–7.0.
- Spraying SSPS solution over noodles increases their water-holding capacity, thus producing an elastic texture.
- For foam stabilization, ~0.2% SSPS may be sufficient or at least beneficial.

References and Further Reading

Anon. 1971. Develops milk-orange juice. *Food Eng.* April:97–101.

Chen, W., Duizer, L., Corredig, M., and H. Douglas Goff. 2010. Addition of soluble soybean polysaccharides to dairy products as a source of dietary fiber. *J. Food Sci.* 75:C478–84.

Furuta, H., and H. Maeda. 1999. Rheological properties of water-soluble soybean polysaccharides extracted under acidic conditions. *Food Hydrocolloids* 13:267–74.

Furuta, H., Nakamura, A., Ashida, H., Asano, H., Maeda, H., and T. Mori. 2003. Properties of rice cooked with commercial water-soluble polysaccharides extracted under weakly acidic conditions from soybean cotyledons. *Biosci. Biotechnol. Biochem.* 67:677–83.

Glahn, P. E. 1982. Hydrocolloid stabilization of protein suspensions at low pH. In *Progress in food & nutrition science*, vol. 6, eds. G. O. Phillips, D. J. Wedlock, and P. A. Williams, 171–7. Oxford: Pergamon Press.

Hojjati, M., Razavi, S. H., Rezaei, K., and K. Gilani. 2011. Spray drying microencapsulation of natural canthaxantin using soluble soybean polysaccharide as a carrier. *Food Sci. Biotechnol.* 20:63–9.

Hou, Z., Gao, Y., Yuan, F., Liu, Y., Li, C., and D. Xu. 2010. Investigation into the physicochemical stability and rheological properties of β-carotene emulsion stabilized by soybean soluble polysaccharides and chitosan. *J. Agric. Food Chem.* 58:8604–11.

Hou, Z., Zhang, M., Liu, B., Yan, Q., Yuan, F., Xu, D., and Y. Gao. 2012. Effect of chitosan molecular weight on the stability and rheological properties of β-carotene emulsions stabilized by soybean soluble polysaccharides. *Food Hydrocolloids* 26:205–11.

Kawamura, S., Kobayashi, T., Osima, M., and M. Mino. 1955. Studies on the carbohydrate of soybeans. *Bull. Agric. Chem. Soc.* 19:69–76.

Liu, J., Verespej, E., Alexander, M., and M. Corredig. 2007. Comparison on the effect of high-methoxyl pectin or soybean-soluble polysaccharide on the stability of sodium caseinate-stabilized oil/water emulsions. *J. Agric. Food Chem.* 55:6270–8.

Luck, H., and J. Grothe. 1973. Fruit-juice-flavored milk. *S. Afr. J. Dairy Technol.* 5:47–52.

Maeda, H. 1992. Development and application of soybean polysaccharide. *Shokuhin to Kaihatsu (in Japanese)* 27:47–9.

Maeda, H. 1994. Soluble soybean polysaccharide: Properties and application of SOYAFIBER-S. *Food Industry (in Japanese)* 37:71–4.

Maeda, H., and A. Nakamura. 2009. Soluble soybean polysaccharide. In *Handbook of hydrocolloids*, eds. G. O. Phillips and P. A. Williams, 693–709. Boca Raton, FL: CRC and Oxford: Woodhead Publishing Limited.

Nakamura, A., Furuta, H., Maeda, H., Nagamatsu, Y., and A. Yoshimoto. 2000. The structure of soluble soybean polysaccharide. In *Hydrocolloids: Part 1*, ed. K. Nishinari, 235–41. Amsterdam: Elsevier Science.

Nakamura, A., Furuta, H., Maeda, H., Takao, T., and Y. Nagamatsu. 2002. Analysis of the molecular construction of xylogalacturonan isolated from soluble soybean polysaccharide. *Biosci. Biotechnol. Biochem.* 66:1155–8.

Nakamura, A., Takahashi, T., Yoshida, R., Maeda, H., and M. Corredig. 2006. Emulsifying properties of soybean soluble polysaccharide. *Food Hydrocolloids* 18:795–803.

Nakamura, A., Yoshida, R., Maeda, H., and M. Corredig. 2006. The stabilizing behaviour of soybean soluble polysaccharide and pectin in acidified milk beverages. *Int. Dairy J.* 16:361–9.

Nakamura, A., Yoshida, R., Maeda, H., Furuta, H., and M. Corredig. 2004. A study of the role of the carbohydrate and protein moieties of soy soluble polysaccharides in their emulsifying properties. *J. Agric. Food. Chem.* 52:5506–12.

Nussinovitch, A. 1997. *Hydrocolloid applications. Gum technology in the food and other industries.* London: Blackie Academic & Professional.

Nussinovitch, A. 2003. *Water soluble polymer applications in food.* Oxford, UK: Blackwell Publishing.

Nussinovitch, A. 2010. *Plant gum exudates of the world sources, distribution, properties and applications.* Boca Raton, FL: CRC Press, Taylor & Francis Group.

Nussinovitch, A., and M. Hirashima. 2014. *Cooking innovations, using hydrocolloids for thickening, gelling and emulsification*. Boca Raton, FL: CRC Press, Taylor & Francis Group.

Shurtleff, W., and A. Aoyagi. 1984. *Tofu and soymilk production*. Lafayette, CA: The Soyfood Center.

Tan, C. T., and J. W. Holmes. 1988. Stability of beverage flavor emulsions. *Perfumer and Flavorist*. Feb/Mar: 23–41.

Yoshii, H., Furuta, T., Maeda, H., and H. Mori. 1996. Hydrolysis kinetics of okara and characterization of its water-soluble polysaccharides. *Biosci. Biotechnol. Biochem*. 60:1406–9.

CHAPTER 18

Vegetable Protein Isolates

18.1 INTRODUCTION

New protein supplies are required to support the dietary needs of the growing population. Vegetable proteins are becoming more and more important for numerous reasons, and such reasons are related to market price, land requirement, and food safety, among many others. The major resources for vegetable proteins are legume grains, cereals, oilseeds, root vegetables, leaves from land or aquatic plants and fruits.

18.2 MAIN SOURCES

18.2.1 LEGUMES

Legume grains include peas, soybeans, chickpeas, and peanuts. Legumes have been used for food purposes for more than 20,000 years. They represent crops from the family Leguminosae. Generally, legume plants can fix atmospheric nitrogen to fabricate their own protein compounds as a result of symbiotic associations with rhizobia (i.e., nitrogen-fixing bacteria) present in the root nodules. Farmed legumes consist of forage and grain legumes. The first group is mainly grown for hay or grazing, and the second for their grains, which are marketed as dry products. Legume foliage and grains have high protein content and are, therefore, of interest to agriculture. Major grain legumes are peas, faba beans, lentils, soybeans, lupins, and chickpeas. About 75% of the grain legumes that are produced worldwide are soybeans. In the last 30 years, soybean production has increased over 2-fold in comparison to ~50% for other grain legumes. Soybean is a universal crop. It presumably originated in China thousands of years ago, and was introduced in Europe at the beginning of the 18th century and in the United States in the late 18th century. Today, the U.S. produces half of the total world production. Pea is native to Southwest Asia and was among the initial crops cultivated by man. Pea was introduced in the U.S. some time after the 15th century. It is regarded as a crop that grows in the cold seasons, and it is grown in climates that can be found, for example, in Northern Europe. Cereals were probably farmed thousands of years ago. The first recorded farming communities date back to 9000 BC, in an area between the Tigris and Euphrates Rivers. Rice, barley, and wheat were grown in ~2000 BC in North India. Cereals are cultivated in regions with semiarid to very hot or cold conditions. The main cereal crops are wheat, rice, maize, barley, and sorghum. Wheat has been used as food since early human history, and it is among the oldest and largest comprehensively grown crops. Evidence for the cultivation of wheat in different parts of the world dates back thousands of years, the oldest record being in the Nile valley in ~5000 BC. From the Mediterranean, cultivation expanded to Asia and Europe. Wheat can be grown in different climates and the cultivars can differ in their kernel properties, grain hues, and protein contents.

18.2.2 Oilseeds

Oilseeds are grown mostly for edible oil production. Oilseeds include sunflower, rapeseed, sesame, cotton seed, and safflower. In a wider sense, soybeans and peanuts may also be considered as oilseeds. Sesame and rapeseed oilseeds were included in Indian Sanskrit scripts from 2000 BC. Sunflower was described in present-day New Mexico and Arizona in ~3000 BC. After oil extraction, the remaining cake acts as a rich source of proteins. Peanut and sunflower have 50%–60% and 40%–50% proteins, respectively. Other oilseed cakes contain ~35%–45% proteins. It is thus not surprising that cottonseed, peanut, rapeseed, soybean, and sunflower are used in significant proportions for protein meal production.

18.2.3 Root Vegetables

Root vegetables include a broad variety of roots and tubers. Some known root vegetables include potato, cassava, and sweet potato. They have been important parts of the human diet from a very early stage of human evolution. Once their cultivation began, roots and tuber crops became irreplaceable food sources. Root vegetables are a good source of carbohydrates, but their protein content is not noteworthy. Potato has been cultivated for thousands of years in South America. It was introduced in Europe in the second half of the 16th century, and is currently the fourth major crop after wheat, rice, and maize. After starch production, we are left with fibers and "potato juice". The latter contains the proteins; nevertheless, it comprises a small fraction of the total worth of commercial potato crops as a result of their poor solubility after the industrial process.

18.2.4 Green Leaves and Fruits

Leaves from alfalfa, cassava, amaranth and aquatic plants, and seeds from fruits such as grapes, tomatoes, and papayas, may contain proteins that can be used for human consumption. Leaf protein concentrate serves as an important protein supplement, but its bitter and grassy flavor and the dark green color need to be eliminated before use. Also, different fruit byproducts can also be used, such as date seeds, grape seeds, tomato seeds (which contain ~20%–30% protein), papaya seeds, and mango and apricot kernels.

18.3 CHEMICAL COMPOSITION OF VEGETABLE PROTEINS

In general, plants contain numerous chemical compounds. Their character depends on the growing conditions, physiological issues, genotype, and post-harvest conditions. Legumes are a source of complex carbohydrates, protein, and dietary fiber. Legume seeds contain 17%–40% proteins. Soybeans are comprised of ~40% proteins, 35% carbohydrates, and 20% lipids. Different cultivars contain large variations in their contents. Peas contain 20%–25% protein, 33%–50% starch, and low lipid levels. Cereal grains consist of 7%–13% protein. The approximate protein content in wheat, rice, maize, barley, sorghum, rye, and oats is 11%, 7%, 10%, 11%, 8%, 9%, and 9%, respectively. It is important to note that the relevant literature and data show large differences in the reported compositions. Proteins and oil are the main components of oilseeds, and ~16%–40% of the weight of the oilseed is protein. The protein content of potatoes ranges from 0.7%–4.6%. In most vegetables, the protein content is ~1%–5%. In fruits, proteins are present in small amounts. Nuts are exceptional because of the presence of ~20% nitrogen compounds.

18.4 PROTEIN COMPOSITION

Proteins are complex macromolecules. Their primary structure consists of a linear sequence of amino acids, but knowing this structure is not sufficient to understand the protein's function. The secondary, tertiary, and quaternary structures of proteins provide information on the actual extent of exposure of hydrophobic regions, about the apparent net charge and the size and shape, among other parameters. Seed proteins can be divided into three groups: storage proteins, biologically active proteins, and structural proteins. In general, a protein is known as a storage protein when it is present at a concentration ≥5%. These proteins are generated during seed development and are stored in the seed to be used during germination. Other relevant characteristics are: deficiency of enzymatic activity, occurrence in membrane-bound vesicles, prevalence of a multimeric structure, and polymorphism of their polypeptides. Seed proteins include water-soluble albumins, globulins that are soluble in dilute salt solutions, alcohol-soluble prolamins, and glutelins that can be extracted with weak alkaline, weakly acidic, and dilute detergent solutions. Further information about these storage proteins can be located elsewhere. The most important soybean proteins are glycinin and β-conglycinin. Dry pea seeds enclose

~25% proteins, of which 70%–80% are globulins: legumin, vicilin, and convicilin. Wheat contains 80% gluten as its major protein constituent, while albumins and globulins are present as the minor proteins. The soluble potato tuber proteins fall into three groups: The first two groups are patatin, i.e., the major protein, and the protease inhibitors, while the third group consists of all other proteins.

18.5 MANUFACTURE

Many parameters must be considered when selecting a method for protein isolation. Isolation is done by disrupting the tissues of the raw material that contain the protein and then preparing the extract. Efficiency of the isolation process depends on the solubility of the components, the time–temperature relationships, ionic strength, and pH. For proteins obtained from most plant sources, a pH of 4.0–9.0 is regarded as safe. With insoluble proteins, organic solvents or surfactants need to be used. After the protein is extracted, it needs to be separated. This can be achieved by clarification via centrifugation, filtration, or membrane processes. The clarified protein solution is then concentrated by ultrafiltration or precipitation. The latter can be achieved with isoelectric precipitation, organic solvents, salting-out method, salting-in method (reduction of ionic strength), exclusion polymers, or heating. Following these steps, a protein preparation is obtained that has been purified by several degrees. Further information on the preparation of specific protein isolates can be found elsewhere.

18.6 FUNCTIONAL PROPERTIES

Proteins can help in the production of foams, emulsions, gels, texturized products, and dough, among many other products. Proteins can produce oil-in-water emulsions, but not water-in-oil ones. To homogenize the emulsion, intense blending and/or agitation is required. In this case, the effective molar mass of the protein is the most relevant parameter during emulsion formation; the greater it is, the higher the quantity of protein needed to obtain minute droplets. At elevated protein concentrations, there are no noticeable differences among proteins, with respect to their ability to form small droplets and stabilize the emulsion. Proteins can serve as appropriate surfactants to produce foams, especially the ones prepared by beating, but there are great differences in the foaming abilities of different proteins. This is due to the considerable conformational stability of a number of proteins that, over short time scales, can barely adsorb to the air–water interface. The role of proteins in foams is to augment the viscosity and elastic properties of the liquid, while decreasing the interfacial tension and producing strong films.

Protein gels can be used in many industrial applications. Although gelatin is very popular, there is a demand for, and increasing interest, in vegetable-based products. Gel formation often involves steps such as denaturation, aggregation, and network formation. Protein denaturation is commonly achieved by heating. Gelation can be induced by high hydrostatic pressure, enzymes, or via cold gelation. In the process of denaturation the unfolded conformation of the protein is attained. In this step, functional groups become exposed, which can then interact to form aggregates that may gel at high protein concentrations. Fine-stranded and coarse gel networks can be formed. The coarse gels are produced by random aggregation, whereas the fine-stranded gel proteins are attached to each other forming a string. The type of gel produced also depends on the pH; the closer the pH is to the isoelectric point and the higher the ionic strength is, the better is the prospect of getting coarser gels. The 3D structures of proteins lead to many unique products, such as meat analogues and fat substitutes. Texturization involves protein denaturation, followed by reorganization and some extent of orientation of completely unfolded proteins. This can be achieved by surface precipitation, rolling/spreading, or extrusion. The food industry uses extrusion cooking, spinning, enzymatic or chemical texturization, freeze texturization, or high-pressure texturization to achieve these goals. Proteins can also interact with water and bind lipids. Proteins that contain large amounts of charged amino acids tend to bind large amounts of water. Binding lipids might improve flavor retention and mouthfeel. Therefore, it is not surprising that protein isolates are customarily included in ground meat products to decrease the cooking losses that result from fat separation. Upon dough production and during dough mixing, gliadins and glutenins unfold, exposing the hydrophobic groups that cause protein aggregation and cross-linking due to the formation of disulfidic groups. The result is a viscoelastic network structure that encloses the starch granules and expands when gas is formed by yeasts or when water vapor is formed during heating.

18.7 REQUIRED FUNCTIONAL PROPERTIES FOR FOOD APPLICATIONS

Food applications involving bakery products require viscosity, elasticity, gelation, and water-binding properties. These properties are achieved via the protein characteristics/mechanisms of hydrophilicity, disulfide cross-linking, network formation and hydrogen bonding. An example of a product having these properties is cereals. In low-fat baked products, fat and flavor binding are the most important properties, and these can be obtained through hydrophobic bonding entrapment. In desserts and dressings, the required properties are solubility, emulsifying/foaming properties, and fat mimetics. The main protein sources for achieving these properties are soybean, peanut, lupin, other legumes, and oilseeds. The protein requirements are hydrophilicity, molecular flexibility, and interfacial adsorption capacity. For dairy substitutes, the required properties are solubility, absence of color or taste, emulsification properties, and heat stability. The protein requirements are the same as that for desserts, and the relevant protein sources are the same as well. Infant formulas require high nutritional value as well as solubility, emulsifying properties, and heat stability. This can be achieved by using soybean proteins as well as vegetable protein hydrolysates. In this case, high digestibility, total absence of allergens, hydrophilicity, and interfacial adsorption are the requirements from the proteins. For beverages, soups, and gravies, the required properties include solubility, viscosity, and acid stability and these can be achieved by using proteins from sources such as soybean, pea, vegetable protein hydrolysates, fermented cereals, and legumes. Another interesting food application is in cheese-like products. For these, the required properties include gelation, solubility, and absence of color and taste. The protein requirement/mechanism includes hydrophilicity, protein solvation, network formation, water entrapment, and immobilization and these requirements can be met by soybean. The above paragraph presents only a few food applications, but numerous other food examples can be located elsewhere.

18.8 FOOD APPLICATIONS IN FOOD PRODUCTS

There is a major market for protein preparations obtained from soybean. Other grains have also attracted much interest. In the Far East, traditional food items, such as *tofu*, soymilk, and fermented products are extensively consumed. In the West, the use of soybean products is on the rise and refined soybean protein ingredients are used to a large extent. Defatted soybean meal is used mostly in animal feed. Limitation in the use of soybean is associated with the presence of lipoxygenases, which oxidize fatty acids and may cause rancidity in the food. The presence of anti-nutritional factors can be overcome by proper heating. Today, there is a long list of soybean protein preparations that are available in the market. These can be applied in numerous food products, and their contribution as functional and nutritional substitutes for animal proteins is very important. These substitutes can have meat-like manifestations, with the desired texture, appearance, and cooking performance.

Both soluble soybean polysaccharide (SSPS) (Chapter 17) and high-methoxyl pectin can stabilize casein particles in low-pH milk. SSPS seems to balance pectin, as at pH 4.6, pectin stabilizes acid dispersions, while under acidic conditions (pH 3.2–4.0), SSPS adsorbs onto the surface of casein micelles (Chapter 14) and prevents aggregation. The dissimilarities in the functions of these two polysaccharides are caused by the difference in their molecular structures. SSPS-fortified dairy products (beverages, puddings, and low-fat ice cream) were acceptable to consumers since these products maintained the textural range of the existing products. Based on rheological and consumer sensory testing, 4% SSPS in a dairy beverage, 4% SSPS in a pudding with κ-carrageenan, and 2% SSPS in low-fat ice cream were the most acceptable levels of SSPS incorporation. However, the levels of soluble fiber added in these new products are considerably higher than in the products currently available in the market, so SSPS-fortified products can help consumers increase their soluble dietary fiber intake.

SSPS can be used for microencapsulation purposes. For instance, microencapsulated canthaxanthin was manufactured by using SSPS as a wall material via the spray-drying technique. The SSPS demonstrated very good microencapsulation ability owing to its emulsification properties. The best ratio of core-to-wall was 0.25:1, since the microcapsules prepared with this ratio had the smallest-sized droplets and microcapsules; in addition, they showed the highest microencapsulation efficiency and the lowest loss during the process. SSPS can also be used as a stabilizer in different emulsions. Stable oil-in-water emulsions incorporating β-carotene in oil droplets that were surrounded by multiple layers of interfacial membranes were produced using a two-stage process of homogenization, which relied on the adsorption of chitosan (Chapter 5) to anionic droplets coated with SSPS. Results showed that the viscoelastic performance could be improved by the adsorption of chitosan onto the SSPS-coated droplet surfaces. Emulsions prepared with SSPS are stable in a pH range of 3.0–7.0. The emulsifying properties of SSPS are not affected by pH or ionic strength. When compared to existing data on the emulsification properties of gum arabic or modified starch, SSPS is found to present similar behavior in acidic, concentrated, or dilute emulsions. However, lower amounts of SSPS are needed to stabilize acidic emulsions as compared to those reported for gum arabic and modified starch.

Soybean proteins might be included in bakery products to enhance their nutritional value and shelf life. Pea meal consists mainly of starch. Textured pea protein can be used in vegetarian food such as vegetarian sausages and burgers. Ingredients from pea protein preparations, such as arrum (prepared from yellow peas and wheat gluten in a ratio of 1:1), have been used for the production of artificial meat chunks and the ingredients of lasagna, burgers, and pies. Various lupin protein products are available for use in the food industry. Several such products can be found in the baking industry, for example, emulsified products and extruded food. Gluten is processed and extruded to mimic the texture of meat. Oilseed protein-based products are also produced, but to a lesser extent. A few examples are infant formulas, bakery and pasta products, and milk substitutes; other uses are also being explored to tap the recent improvements in the flavor, color, and functionality of these proteins. With respect to bakery and pasta products, the gelatinization and retrogradation behaviors of wheat starch in an aqueous system were investigated, in the presence or absence of some anionic polysaccharides, SSPS, and gum arabic. The addition of each polysaccharide (0.1%–1% w/v) during gelatinization decreased the peak viscosity of the composite system (starch concentration: 5% or 13%), and this effect was greater for SSPS than for gum arabic, at higher starch concentration. The addition of each polysaccharide (0.1%–1%) decreased the amount of amylase leached during gelatinization, and this effect was also generally greater for SSPS (Chapter 17) than for gum arabic. The quality of potato proteins is fairly good, but they have a major anti-nutritional compound, glycoalkaloids. Moreover, the presence of protease inhibitors is unwelcome in food. Therefore, research efforts are aimed at overcoming extensive protein insolubility. Potato proteins can be included in baked goods, especially crispbread, where the protein content can be doubled without negatively influencing its texture or structure.

18.9 REGULATORY STATUS

Vegetable protein products (VPPs) are produced from different vegetable sources and their use in food is regulated to a considerable extent. The regulation covers crucial aspects, such as product labeling, the use of genetically modified organisms (GMOs), the introduction of novel food products, food allergens, and the modification of food products. A label must indicate any presence of VPPs in the food. The label must also be in accordance with general standards. The vegetable source must be stated (for instance, soybean, wheat, pea, etc.) accompanied by a complete list of included ingredients. The level of VPP should be calculated on a dry weight basis when low levels are used for functional purposes. The protein's nutritional values should be assessed and, most importantly, the corresponding protein should contain a minimal percentage of ingredients such as lysine, methionine, cysteine, or tryptophan, especially if the diets are deficient in those. When VPP is used to replace an animal protein completely, the name of the food should carry the name of the VPP with a proper description. In all cases, nutritional value, microbiological, and toxicological aspects of the VPP should be examined. Some VPPs also have specific Codex standards. If GMO material is to be used, the company should have it approved to avoid adverse effects on human health.

Some protein-containing food items may cause allergies in humans. In case of such vegetable products, the label should state whether these proteins are used in their original or derived form. Finally, standards have been fixed for using contaminants, such as chloropropanols, with toxicological properties. The relevant food industry should modify the production process to prevent the occurrence of such undesirable flavor components.

18.10 RECIPES

18.10.1 *MOMEN TOFU* (SOYBEAN CURD) (FIGURE 18.1)

(Makes 600 g)
500 g (3 cups) soybeans
2 L water (2 times the volume of swollen soybeans)
80–100 g (⅓–⅖ cups) bittern (4%–5% soybean milk)

1. Soak soybeans in water overnight (Hint 18.10.1.1).
2. Put the swollen soybeans (1) and 1 L water (Figure 18.1A) in a food processor or blender and pulverize (Figure 18.1B) (Hint 18.10.1.2).
3. Add remaining water to the soybean paste (2) and heat in a pan while stirring until it boils (90°C) (Figure 18.1C) (Hint 18.10.1.3). Continue to boil for a few minutes.

FIGURE 18.1 *MOMEN TOFU* (SOYBEAN CURD). (A) PUT SWOLLEN SOYBEANS AND WATER IN A FOOD PROCESSOR. (B) PULVERIZE. (C) ADD WATER TO SOYBEAN PASTE AND HEAT WITH STIRRING UNTIL IT BOILS. (D) STRAIN SOYBEAN MILK FROM SOYBEAN FIBER. (E) SQUEEZE OUT THE HOT SOYBEAN MILK. (F) PUT BITTERN IN A BOWL. (G) POUR SOYBEAN MILK OVER BITTERN. (H) WHEN THE MIXTURE BEGINS TO COAGULATE, LADLE OUT THE SOYBEAN CURD. (I) POUR INTO MOLDS AND COVER WITH A PIECE OF CLOTH. (J) PRESS TO REMOVE WATER. (K) REMOVE *TOFU* FROM THE MOLDS. (L) PUT IT IN WATER.

4. Separate soybean milk from soybean fiber (*okara*) in a strainer with a piece of cloth (Figure 18.1D). Squeeze the hot soybean milk by hand (Figure 18.1E) (Hint 18.10.1.4).
5. Measure volume of soybean milk (4) and heat to 80°C.
6. Put bittern in a bowl (Figure 18.1F) and rapidly pour in the hot soybean milk (5) (Figure 18.1G); mix gently (Figure 18.1H) and then let it stand for 10–15 minutes.

7. When the soybean curd (6) begins to coagulate, ladle it into molds, cover with cloth (Figure 18.1I), and press down the lid to remove water (Figure 18.1J).
8. Remove *tofu* (7) from the molds (Figure 18.1K) and put in water (Figure 18.1L).

pH of *tofu* = 5.83

pH of soybean milk = 6.60

pH of bittern = 7.29

Preparation hints:

18.10.1.1 Soak soybeans until they are fully swollen.

18.10.1.2 Add the soybeans in several batches, pulverizing each in turn.

18.10.1.3 Be careful as bubbling will start suddenly when the soybean paste boils.

18.10.1.4 Use rubber gloves because the preparation is very hot.

Momen tofu is very popular in Japan; the name means cotton-like *tofu*. *Tofu* can be processed into many kinds of food, such as thick fried *tofu* (Chapter 5), thin fried *tofu*, grilled *tofu*, etc.

18.10.2 Soybean Milk and Black Sesame Seed Pudding (Figure 18.2)

(Serves 5)

400 g (a little over 1⅗ cups) soybean milk

8 g (a little under 1 tsp) black sesame seeds

100 g (½ cup) brown sugar

100 g (2) eggs

Black sesame seeds

1. Grind black sesame seeds with an earthen mortar and pestle (Figure 18.2A) (Hint 18.10.2.1) to a paste-like consistency.

Figure 18.2 Soybean milk and black sesame seed pudding. (A) Grind black sesame seeds to a paste-like consistency. (B) Heat soybean milk and sugar while stirring. (C) Add lukewarm soybean milk to ground sesame seed paste gradually while stirring. (D) Pour beaten egg into the dispersion of soybean milk while stirring. (E) Pour the strained dispersion into molds and steam on low heat. (F) Garnish with black sesame seeds.

2. Heat soybean milk and sugar in a pan, while stirring with a whisk, to 45°C (Figure 18.2B).
3. Add the lukewarm soybean milk (2) to the ground sesame paste (1) gradually while stirring with a whisk (Figure 18.2C).
4. Beat eggs in another bowl and pour into the soybean milk dispersion (3) gradually (Figure 18.2D) while stirring with a whisk.
5. Strain the soybean milk dispersion (4).
6. Pour the strained dispersion (5) into molds and steam the molds on low heat for 17 minutes (Figure 18.2E).
7. Garnish with black sesame seeds (Figure 18.2F).

pH = 6.34

Preparation hints:

18.10.2.1 You can also use a grinder or blender.

This is a milk-free dessert. It is easier to set by steaming than caramel custard with milk. It comes in many variations other than black sesame; for example, with green tea, honey, pumpkin paste, etc.

18.10.3 *Unohana* (Seasoned *Okara*) (Figure 18.3)

(Serves 5)

200 g (a little under $1\frac{1}{3}$ cups) *okara*
50 g carrot
50 g *konjac*
50 g long onion
5 g dried *shiitake* mushroom
150 mL ($\frac{3}{5}$ cup) water
18 g (2 tbsp) white sugar
54 mL (3 tbsp) soy sauce
18 mL (1 tbsp) *mirin*
Vegetable oil, salt

1. Wash dried *shiitake* mushroom and soak in water overnight in the refrigerator (Figure 18.3A) (Hint 18.10.3.1).
2. Cut carrot, konjac, long onion, and swollen *shiitake* (1) into small pieces (Figure 18.3B).

Figure 18.3 *Unohana* (seasoned *okara*). (A) Soak washed, dried *shiitake* mushrooms in water overnight in the refrigerator. (B) Cut carrot, konjac, long onion, and swollen *shiitake* into small pieces. (C) Stir-fry. (D) Add *okara* to the stir-fried vegetables and continue frying. (E) Add the water in which the *shiitake* mushrooms were soaked, sugar, soy sauce, *mirin*, and salt to the stir-fried vegetables and *okara*. (F) Simmer until the mixture dries out.

3. Heat vegetable oil in a pan, stir-fry the cut carrot and long onion (2), add the cut *konjac* and *shiitake* (2), and continue stir-frying (Figure 18.3C).
4. Add *okara* to the stir-fried vegetables (3) and continue to stir-fry until water evaporates and the *okara* becomes dry (Figure 18.3D).
5. Add the water in which the *shiitake* mushrooms were soaked. Then add sugar, soy sauce, *mirin,* and salt to the stir-fried vegetables and *okara* (4) (Figure 18.3E) and simmer to dryness (Figure 18.3F).

pH = 5.62

Preparation hints:

18.10.3.1 A type of Japanese broth can be extracted from dried *shiitake* mushrooms. Save the soaking water for use as broth later in the recipe.

This is a traditional Japanese dish. *Okara* is mostly used as *unohana* in Japan. You can add your favorite ingredients, such as chicken, burdock root, onion, fish paste, etc.

18.10.4 *Okara* Pound Cake (Figure 18.4)

(Makes 1 small pan)

100 g (⅘ cup) *okara*
100 g (⅘ cup) weak flour
5 g (1¼ tsp) baking powder
100 g (2) eggs
60 g (a little over ⅓ cup) white sugar
20 g (a little under 1 tbsp) honey
50 mL (⅕ cup) milk
Baking paper

1. Preheat oven to 180°C and line a pound cake pan with baking paper.

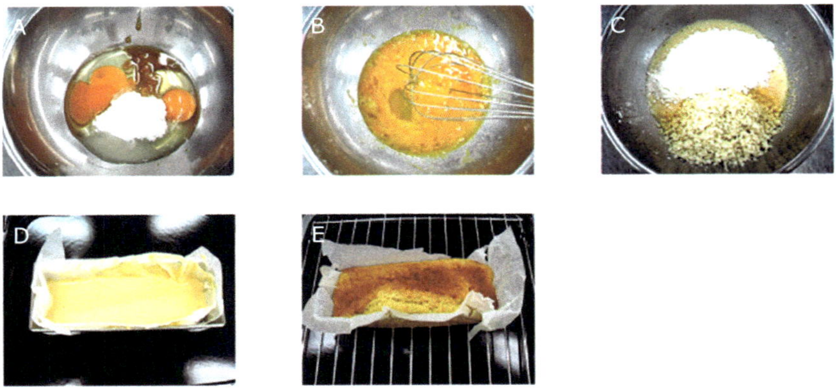

FIGURE 18.4 *OKARA* POUND CAKE. (A) MIX EGG, SUGAR, AND HONEY. (B) ADD MILK TO THE EGG MIXTURE AND MIX. (C) ADD *OKARA*, SIEVED FLOUR, AND BAKING POWDER. (D) POUR THE BATTER INTO THE PAN. (E) COOL THE BAKED CAKE.

2. Sieve together flour and baking powder.
3. Thoroughly mix egg, sugar, and honey in a bowl with a whisk (Figure 18.4A).
4. Mix milk into the egg mixture (3) with a whisk (Figure 18.4B) and add *okara*, the sieved flour, and baking powder (2). Mix with a spatula (Figure 18.4C).
5. Pour the batter (4) into the pound cake pan (1) (Figure 18.4D) and bake at 180°C for 30 minutes.
6. Cool the pound cake on a cake cooler or wire rack (Figure 18.4E) (Hint 18.10.4.1).

pH of batter = 6.69

Preparation hints:

18.10.4.1 Cool and dry thoroughly because this cake contains more moisture than a normal pound cake.

This recipe originated to make use of the *okara* left over as a byproduct of *tofu*. Today it is used to make low-calorie diet cakes.

FIGURE 18.5 BASIL AND SUNFLOWER SEED SAUCE. (A) PLACE ALL INGREDIENTS IN A FOOD PROCESSOR. (B) PURÉE. (C) SERVE WITH PASTA.

18.10.5 BASIL AND SUNFLOWER SEED SAUCE (FIGURE 18.5)

(Serves 5)
30 g (30–35 leaves) basil
120 g (1 cup) sunflower seeds
90 g (½ cup) olive oil
20 g (2 cloves) garlic
Salt

1. Place all ingredients in a food processor (Figure 18.5A) and blend to a purée (Figure 18.5B).

pH = 5.52

This recipe is good on pasta (Figure 18.5C). Pine nuts are also used as a substitute for sunflower seeds.

18.10.6 LOTUS ROOT BALLS (FIGURE 18.6)

(Serves 5)
500 g lotus root
36 g (4 tbsp) potato starch
3 g (½ tsp) salt
15 mL (1 tbsp) *sake*
120 g (12) small shrimp
100 g chicken
Vinegar, potato starch, vegetable oil for deep-frying

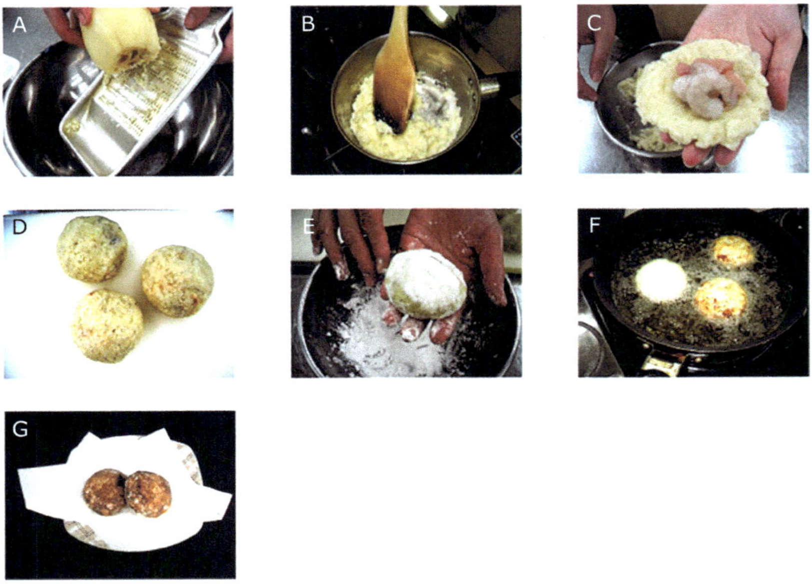

Figure 18.6 Lotus root balls. (A) Grate the strained lotus root. (B) Heat the grated lotus, add potato starch, salt, and sake and simmer until thickened. (C) Put the shrimp and cut chicken in the center of each lotus paste ellipse. (D) Wrap to make balls. (E) Sprinkle potato starch over the balls. (F) Deep-fry. (G) Serve.

1. Peel lotus root and soak it in vinegar and water (Hint 18.10.6.1). Blot the lotus root dry with a paper towel and grate it (Figure 18.6A).
2. Heat the grated lotus root (1) in a pan over low heat, add potato starch, salt, and *sake*, and simmer until thickened (Figure 18.6B). Put it in a bowl and let it stand overnight (Hint 18.10.6.2).
3. Shell the shrimp and cut chicken into 1 cm cubes.
4. Divide lotus root paste (2) into five portions and form each into a flattened ellipse; put the shrimp and cut chicken in the center of each (Figure 18.6C) and wrap them, making balls (Figure 18.6D).
5. Sprinkle potato starch over the balls (4) (Figure 18.6E).
6. Heat vegetable oil for deep-frying to 170°C, deep-fry the balls (Figure 18.6F), and serve (Figure 18.6G).

Preparation hints:

18.10.6.1 The color of the lotus root, which stems from flavonoids, turns white in an acid solution.

18.10.6.2 Longer standing time results in a stickier paste.

This recipe is served in high-class restaurants in Japan. It is enjoyed for its sticky texture.

18.10.7 Alfalfa Stew (Figure 18.7)

(Serves 5)

500 g potato
300 g (1) onion
300 g (2) carrots
10 g (1 clove) garlic
48 g (4 tbsp) vegetable oil
280–300 g alfalfa sprouts
200 g rice
800 mL (3⅕ cups) vegetable broth or water
Salt, pepper

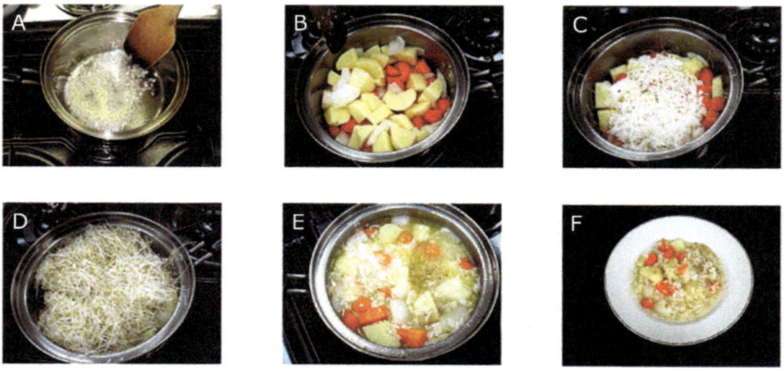

Figure 18.7 Alfalfa stew. (A) Stir-fry chopped garlic. (B) Add cut onion, carrot, and potato to the pot. (C) Add washed rice to the vegetables and continue to stir-fry. (D) Add alfalfa and season with salt and pepper. (E) Add vegetable broth. (F) Serve.

1. Cut onion and carrots into bite-sized pieces and finely chop garlic. Cut potato into bite-sized pieces and soak in water (Hint 18.10.7.1).
2. Heat vegetable oil in a pan, add chopped garlic (Figure 18.7A), and stir-fry it until the vegetable oil takes on a garlicky flavor; add the cut onion, carrot, and potato (1) and continue to stir-fry (Figure 18.7B).
3. Wash rice (Hint 18.10.7.2), add to the stir-fried vegetables (2), and continue to stir-fry (Figure 18.7C).
4. Add alfalfa and season with salt and pepper (Figure 18.7D).
5. Add vegetable broth and boil for 20 minutes (Figure 18.7E) and serve (Figure 18.7F).

pH = 5.82

Preparation hints:

18.10.7.1 Soaking the potatoes removes the surface starch and prevents browning.

18.10.7.2 Washing rice results in a non-sticky texture.

This is a Spanish dish originally called *guiso de alfalfa*. It has a simple taste, and is suited to the early summer season.

18.10.8 Fruit Juice with Alfalfa (Figure 18.8)

(Serves 5)

400 g (2 cups) ice

400 g (3 cups) fresh pineapple chunks

Figure 18.8 Fruit juice with alfalfa. (A) Place all ingredients in a food processor. (B) Pour into glasses and serve.

60 mL (4 tbsp) 100% orange juice
200 g (2) bananas
100 g alfalfa sprouts
Orange flavoring

1. Place all ingredients in a blender (Figure 18.8A) and blend.
2. Pour the juice into glasses (Figure 18.8B).

pH = 4.27

This is a health drink. You can add your favorite fruits and vegetables.

18.11 TIPS FOR THE AMATEUR COOK AND PROFESSIONAL CHEF

- Soy proteins settle in an acid solution because their isoelectric point is at pH 4.0–5.0.
- Soy proteins gelate above 60°C.
- Soy proteins are used in *tofu*, meat, and fish products as a binding agent.
- The emulsifying property of soy proteins is important for sausage production.

REFERENCES AND FURTHER READING

Allen, G. 1999. *Protein: A comprehensive treatise.* Stamford, CT: JAI Press Inc.
Altschul, A. M., and H. L. Wilcke. 1985. *New protein foods.* Orlando: Academic Press Inc.
Burton, W. G. 1969. *Potato, Encyclopedia Britannica.* Chicago: Benton.
Chen, W., Duizer, L., Corredig, M., and H. D. Goff. 2010. Addition of soluble soybean polysaccharides to dairy products as a source of dietary fiber. *J. Food Sci.* 75:478–84.
Creighton, T. E. 1996. *Proteins: Structures and molecular properties.* New York: W.H. Freeman and Co.
Damodaran, S., and A. Paraf. 1997. *Food proteins and their applications.* New York: Marcel Dekker Inc.

Daussant, J., Mosse, J., and J. Vaughan. 1983. *Seed proteins*. London: Academic Press.
Dendy, D. A. V., and B. J. Drobraszczyck. 2001. *Cereals and cereal products*. Aspen: Springer.
Endres, J. G. 2001. *Soy protein products: Characteristics, nutritional aspects, and utilization*. Champaign, IL: AOCS Press.
Funami, T., Nakauma, M., Noda, S., Ishihara, S., Asai, I., Inouchi, N., and K. Nishinari. 2008. Effects of some anionic polysaccharides on the gelatinization and retrogradation behaviors of wheat starch: Soybean-soluble polysaccharide and gum arabic. *Food Hydrocolloids* 22:1528–40.
Gonzalez-Perez, S., and J. B. Arellano. 2009. Vegetable protein isolates. In *Handbook of Hydrocolloids*, 2nd edn., eds. G. O. Phillips and P. A. Williams, 383–420. Boca Raton, FL: CRC Press, and Oxford: Woodhead Publishing Limited.
Gueguen, J., and Y. Popineau. 1998. *Plant proteins from European crops: Food and non-food applications*. Nantes: INRA Editions.
Hall, G. M. 1996. *Methods of testing protein functionality*. London: Blackie Academic and Professional, an Imprint of Chapman & Hall.
Hojjati, M., Razavi, S. H., Rezaei, K., and K. Gilani. 2011. Spray drying microencapsulation of natural canthaxantin using soluble soybean polysaccharide as a carrier. *Food Sci. Biotechnol.* 20:63–9.
Hou, Z., Gao, Y., Yuan, F., Liu, Y., Li, C., and D. Xu. 2010. Investigation into the physicochemical stability and rheological properties of β-carotene emulsion stabilized by soybean soluble polysaccharides and chitosan. *J. Agric. Food Chem.* 58:8604–11.
Lewis, G., Schrire, B., Mackinder, B., and M. Lock. 2005. *Legumes of the world*. Richmond: The Royal Botanic Gardens, Kew.
Murphy, D. J. 1994. *Designer oil crops: Breeding, processing and biotechnology*. New York: VCH Weinheim.
Nagano, T. 2007. Soy proteins. In *Food hydrocolloids: Development and applications*, ed. K. Nishinari, 330–339. Tokyo: CMC Publications (in Japanese).
Nakamura, A., Takahashi, T., Yoshida, R., Maeda, H., and M. Corredig. 2004. Emulsifying properties of soybean soluble polysaccharide. *Food Hydrocolloids* 18:795–803.
Nakamura, A., Yoshida, R., Maeda, H., and M. Corredig. 2006. The stabilizing behavior of soybean soluble polysaccharide and pectin in acidified milk beverages. *Int. Dairy J.* 16:361–9.
Nussinovitch, A. 1997. *Hydrocolloid applications: Gum technology in the food and other industries*. London: Blackie Academic & Professional.

Nussinovitch, A. 2003. *Water soluble polymer applications in foods*. Oxford: Blackwell Publishing.

Nussinovitch, A., and M. Hirashima. 2014. *Cooking innovations, using hydrocolloids for thickening, gelling and emulsification*. Boca Raton, FL: CRC Press, Taylor & Francis Group.

Osbourne, T. B. 1924. *The vegetable proteins*. London: Longmans, Green and Co.

Riaz, M. N. 2005. *Soy applications in food*. Boca Raton: CRC Press.

Robbelen, G., Downey, R. K., and A. Ashri. 1989. *Oil crops of the world*. New York: McGraw-Hill Publ. Co.

Salunkhe, D. K., Chavan, J. K., Adsule, R. N., and S. S. Kadam. 1992. *World oilseeds: Chemistry, technology and utilization*. New York: Van Nostrand Reinhold.

Shewry, P. R., and R. Casey. 1983. *Seed proteins*. Dordrecht: Kluwer Academic Publishers.

Talburt, W. F., and O. Smith. 1987. *Potato processing*. New York: Van Nostrand.

Yada, R. Y. 2004. *Proteins in food processing*. Cambridge: Woodhead Publishing Limited.

CHAPTER 19

Xyloglucan

19.1 INTRODUCTION

Xyloglucan is a storage polysaccharide found in the seeds of a tamarind tree (*Tamarindus indica*). In addition, xyloglucan is the main structural polysaccharide found in the primary cell walls of higher plants. Tamarind seed xyloglucan (TSX) has different uses in food processing and other industries. Aqueous solutions of TSX have unique applications because they are heat-resistant and are stable to the effects of shear and pH. The hydrocolloid can be utilized as a thickener, stabilizer, gelling agent, ice-crystal stabilizer, and starch modifier. Other potential uses of the gum are in sauces as a thickener, in ice creams, dressings, and flour products.

19.2 ORIGIN, DISTRIBUTION, AND PREPARATION

Xyloglucans are found in various botanical sources. A few examples are dicotyledons, monocotyledons, and gymnosperms. Xyloglucan can make up 20%–30% of the dry weight of the primary cell wall. In addition, the seeds of a few plants have plentiful deposits of xyloglucans adjacent to their cotyledon cells. Although some of the xyloglucan is extractable from the vegetative tissue in cold water, most of it is extractable from the cell wall by alkali. In higher plants, xyloglucans have been found on and between the cellulose microfibrils, and this suggests a potential linkage between xyloglucan and cellulose microfibrils. These linkages probably provide elasticity to the formed structures, which is required by the microfibrils for sliding purposes. It has been hypothesized that cell growth requires the enzymatic cleavage of cross-linked xyloglucan and the release of oligosaccharides. The localization of xyloglucan has been assessed by advanced biological methods. The level of xyloglucan per fresh weight of tissue decreases gradually during growth. Fruits, vegetables, and edible foods may include various quantities of xyloglucan oligosaccharides and polysaccharides. To prepare the hydrocolloid, the tamarind seeds are first washed with water and then heated to make the seed coating friable and brittle. The seeds are decorticated and the endosperm is crushed and ground to yield tamarind kernel powder. The kernel powder is then boiled and agitated with water at ~30–40 times its weight for 10 minutes. It is then left in a settling tank overnight to permit fiber and protein precipitation and settling out. The supernatant fluid is concentrated, mixed with filter-aid, and filtered. The resultant fluid is dried and pulverized to yield the seed extract. Further purification is also possible. The purified gum was first sold on a commercial basis in Japan in 1964.

19.3 PROPERTIES

The chemical structure of xyloglucan depends on the plant from which it is extracted. TSX has a β-(1,4)-linked D-glucan backbone substituted to some extent with α-D-xylopyranose at the O-6 position of its glucopyranosyl residues. TSX has mainly four types of oligosaccharides as the repeating units. Further information about the chemical structure of the repeating units can be found elsewhere. Different molecular weights have been reported for TSX, depending upon the method of determination and the procedure of sample preparation. The cell wall xyloglucan has a lower molecular weight than TSX. Aqueous solutions of TSX demonstrate the typical flow behavior of polymer solutions. At low shear rates, a range of Newtonian viscosity is demonstrated, while at high shear rates, shear thinning is seen.

19.4 TSX INTERACTIONS

At high concentrations, TSX forms a viscous solution in water. A thermoreversible gel can be formed by adding gellan gum, xanthan gum, or some small molecules, such as congo red or catechin, to the TSX solution.

In the presence of sugars, TSX increases viscosity and at 40%–70% sugar, it forms a gel. There are several reports of TSX aqueous solutions forming gels with the addition of ethanol. A gel with alcohol is harder and has a lower melting point than a gel with sugar. TSX sugar gels become harder and more elastic upon passing through a freeze-thaw process. The sol–gel transition of TSX in 15% ethanol takes place at 47°C. The gel structures formed in the presence of different monohydric or polyhydric alcohols are different and this, presumably, depends upon the network and type of associations involved. When <20% alcohol is added to the TSX solution in the presence of sugar, the amount of sugar required for gel formation decreases. Xyloglucan can also form a gel by the enzymatic removal of a part of the galactose residues. When 35% of the galactose residues are removed from the xyloglucan, in a temperature range of 40°C–80°C, a thermoreversible gel is formed from the solution and the solution can turn into a sol at lower as well as higher temperatures. Only TSX has such remarkable attributes.

Xyloglucan can form a thermoreversible gel when its aqueous solution is mixed with polyphenols obtained from green-tea extract. In this case, the gel is created by random aggregation and this process is comparable to the method by which the mixture gels with alcohol. TSX does not demonstrate notable synergy with other hydrocolloids; nevertheless, it was reported that two helix-forming polysaccharides, gellan gum and xanthan gum, can form a complex with TSX, thus resulting in increased viscosity or gel formation. This occurs under conditions when gel formation is not possible by the individual polysaccharides alone. Upon addition of potassium iodide to a TSX solution, a light blue or purple color appears. It is similar to the blue stain that is produced when the same reagents are added to starch. Consequently, TSXs in combination with other xyloglucans or starch are termed as "amyloids". This observation probably results from the formation of a complex of iodine molecules and xyloglucan.

19.5 FOOD APPLICATIONS IN THE FOOD INDUSTRY AND REGULATORY STATUS

In Japan, TSX is widely used as a food additive. It can be used in dressings, flour products, ice creams, and sauces. TSX is utilized as a viscosity former. It uniqueness lies in its ability to create a non-sticky texture. TSX has been utilized at low levels (~0.05%) in meat sauces since they require a low pH, high viscosity, and high stability.

TSX is utilized in fruit beverages, low-fat milk, and cocoa. In fruit-pulp beverages, the small particles within the medium need to be stabilized and TSX might be helpful in this regard. Similar stabilization is achieved in *shiruko* (sweet red bean soup with rice cake). A stable viscosity can be achieved with TSX in batter mixes for fried products, especially during the battering process. The ability of TSX to increase viscosity can be utilized to stabilize salad dressings and mayonnaise. TSX provides more support to smaller oil droplets in emulsion than xanthan or carrageenan does. In general, it is agreed that xanthan has a stronger stabilizing effect than TSX. The stabilization depends upon both concentration and pH. Xyloglucan–protein conjugates demonstrate enhanced emulsifying properties. In addition, the conjugation also contributes to heat stability of the protein. With soy-protein polysaccharide conjugates, xyloglucan conjugate has a better emulsifying effect than conjugates with other polysaccharides. TSX, either by itself or in combination with guar gum, locust bean gum, or carrageenan, is a valuable stabilizer in ice cream and sherbet. When TSX is used in frozen desserts, many advantages—good overrun, impressive heat-shock stability, good water-holding, no separation of sugar and ice crystals—are observed. As a gelling agent, TSX produces an elastic gel in the presence of concentrated sugar solutions. Such a gel is also resistant to the freeze-thaw cycle. The blending of oxidized xyloglucan with chitosan (Chapter 5) forms versatile, irreversible, transparent hydrogels. This complex gel possesses better thermal properties as compared to the native xyloglucan. Incorporation of flavors and salts does not alter its gel nature. In addition, its antimicrobial and textural properties are very promising, and this non-toxic, renewable gel can potentially find extensive use in food applications. TSX can be utilized as a substitute for pectinin pectin jellies or jams and can be included in Japanese traditional desserts, such as *yokan* and *kudzu-mochi*. TSX might suppress the ageing of starch under specific conditions.

As a starch substitute or in combination with starch, TSX can be used in custard cream, flour paste, noodles, and Japanese confections. Partial substitution of tapioca starch with xyloglucan yields highly viscous and pseudoplastic liquids. Xyloglucan appears to form a continuous phase in a tapioca starch/xyloglucan mixture, while it confines starch polysaccharides into discontinuously dispersed phases. As a result, the rheological stability of tapioca starch paste against shear deformation, and during storage, was improved by the partial substitution of tapioca starch with xyloglucan. TSX solution can be used as a fat replacement, except for frying and nutritional applications. Low-fat products, such as mayonnaise and dressing, can be produced by the inclusion of TSX. Sometimes TSX can be used in blends with xanthan, starch, or dextrin. A reduction of ~90% of its viscosity results in the production of low-viscosity TSX (LVTSX). This material is useful in producing a smooth texture and also for textural modifications. Addition of LVTSX to noodles prevents sticking and improves the texture. The noodles are either dipped in or are sprayed with the LVTSX solution. Similar effects can be achieved for cooked rice. At 0.5% addition, LVTSX prevents water release from vegetables and dressings and, thus, the dressing sticks to the vegetables well. Finally, addition of LVTSX improves the mixing properties of dough and the firmness and volume of baked bread and sponge cakes. The effects of fractionated xyloglucan on the physical properties of dough and on bread quality were also studied. Xyloglucan was fractionated by enzymatic hydrolysis. Four kinds of fractions were obtained. Addition of these fractions increased the stability of the dough and improved the loaf and the softness of bread samples. In particular, fraction WS-XG-D (1%–5%) (i.e., the fraction having an average degree of polymerization of 223) significantly improved various parameters in the final products, such as loaf volume, storage properties, and appearance. It also gave a fine distribution of small gas cells. Its addition at a low level (1%) also showed more improvements as compared to other additives. In Japan and countries in Southeast Asia, TSX is a permitted food additive, whereas in the United States and the European Union, it is not.

19.6 RECIPES

19.6.1 Sponge Cake (Figure 19.1)

(Makes 18 cm diameter cake)
0.5 g (⅙ tsp) TSX powder
40 mL (2⅔ tbsp) water
150 g (3) eggs
90 g (⅔ cup) weak flour
90 g (⅖ cup) granulated sugar
30 g unsalted butter
Jam sauce, heavy cream, icing (optional)
Baking paper, hot water

1. Preheat the oven to 170°C and line a round cake pan with baking paper (Figure 19.1A) (Hint 19.6.1.1).
2. Put water in a bowl and stir in TSX (Figure 19.1B) with a hand mixer.
3. Sieve weak flour (Figure 19.1C), soften/melt butter in a microwave.
4. Put egg and sugar in a bowl, place the bowl over a hot water bath (60°C), and stir it with a hand mixer until sugar dissolves (Figure 19.1D); remove from hot water bath.
5. Continue to mix the egg and sugar (4) with a hand mixer for 6–8 minutes until the mixture thickens (Figure 19.1E).

Figure 19.1 Sponge cake. (A) Line a round cake pan with baking paper. (B) Stir TSX into water. (C) Sieve weak flour. (D) Put egg and sugar in a bowl placed over hot water and stir until sugar dissolves. (E) Continue to mix the egg and sugar until the mixture thickens. (F) Gradually add the sieved flour to the thickened egg mixture. (G) Add the TSX solution to the batter. (H) Add softened/melted butter to the batter. (I) Pour batter into the lined cake pan. (J) Bake and cool the sponge cake. (K) Spread jam sauce and whipped heavy cream on the center of the cake and garnish with icing (optional).

6. Add the sieved weak flour (3) to the thickened egg mixture (5) in a few steps, while stirring with a spatula (Figure 19.1F) (Hint 19.6.1.2).
7. Pour the TSX solution (2) (Figure 19.1G) and softened melted butter (3) (Figure 19.1H) into the batter (6), pour into the round cake

pan (1) (Figure 19.1I) (Hint 19.6.1.3), and bake it at 170°C for 25 minutes.

8. Cool the sponge cake on a cake cooler or wire rack (Figure 19.1J).
9. Spread jam sauce (mixture of jam and Kirshwasser or rum) and whipped heavy cream on the center of the cake and garnish with icing (optional) (Figure 19.1K).

pH of batter = 7.80

Preparation hints:

19.6.1.1 Use butter or margarine to stick baking paper to the cake pan.

19.6.1.2 To make a softer cake, it is important not to over-stir; this also prevents an overly sticky batter.

19.6.1.3 Drop the cake pan, containing the batter, on a flat surface a few times to remove the air in the batter.

This recipe makes a kind of classic sponge cake, and can be adapted to taste. Adding TSX to the cake results in a fluffy texture and inhibits starch retrogradation.

19.6.2 *TSUKUDANI* (LAVER PRESERVES) (FIGURE 19.2)

(Serves 5)

30 g roasted laver (Figure 19.2A)
2 g (⅔ tsp) TSX powder
50 mL (a little under 3 tbsp) *mirin**
50 g (a little over ⅓ cup) white sugar
70–100 mL (¼–⅖ cup) soy sauce
300 mL (1⅕ cup) water

1. Put water in a pan and stir in TSX with a hand mixer (Figure 19.2B).
2. Add *mirin*, white sugar, and soy sauce to the TSX solution (1) and mix it with a spatula; add roasted laver, tearing it by hand into small pieces (Hint 19.6.2.1) (Figure 19.2C).

* You can use honey in place of *mirin*.

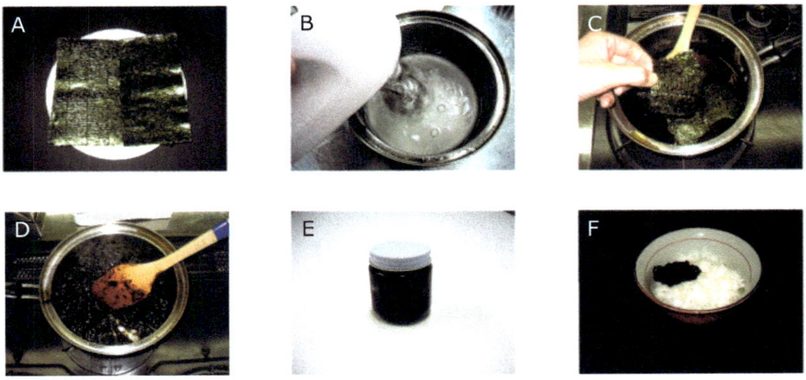

FIGURE 19.2 *TSUKUDANI* (LAVER PRESERVES). (A) ROASTED LAVER. (B) ADD TSX TO WATER AND STIR. (C) ADD SMALL PIECES OF ROASTED LAVER TO THE TSX SOLUTION, ALONG WITH *MIRIN*, WHITE SUGAR, AND SOY SAUCE. (D) SIMMER UNTIL IT BECOMES A PASTE-LIKE JAM. (E) POUR *TSUKUDANI* IN A GLASS CONTAINER. (F) SERVING SUGGESTION: *TSUKUDANI* ON RICE.

3. Heat the pan on low heat and simmer the mixture (2) at 90°C for 20–30 minutes until it becomes a paste-like jam (Figure 19.2D).
4. Pour *tsukudani* into a glass bottle (Figure 19.2E).

pH = 5.26

Preparation hints:

19.6.2.1 Tearing laver into small pieces produces a smooth texture.

This recipe is very popular in Japan, and it is good on rice (Figure 19.2F). Other seafood, such as kelp or small fish, is used to make *tsukudani*, and many kinds of seasonings are added to modify the flavor. TSX is added for its water-holding capacity.

19.6.3 *KUDZU-MOCHI* (JAPANESE ARROWROOT JELLY) (FIGURE 19.3)

(Serves 5)

100 g (5⁄6 cup) *kudzu* (arrowroot) starch (Figure 19.3A)

5 g (1 2⁄3 tsp) TSX powder

50 g (a little over ⅓ cup) white sugar
500 mL (2 cups) water
100 g (⅔ cup) brown sugar
50 mL (3⅓ tbsp) water
Soybean flour

1. Mix *kudzu* starch, TSX, and white sugar well in a pan and then mix in water (Figure 19.3B) (Hint 19.6.3.1).
2. Heat the mixture (1) over medium heat while stirring with a spatula. After the mixture boils, lower the heat and continue to stir (Figure 19.3C) until mixture becomes transparent (at 70°C) (Figure 19.3D) (Hint 19.6.3.2).

FIGURE 19.3 *KUDZU-MOCHI* (JAPANESE ARROWROOT JELLY). (A) *KUDZU* (ARROWROOT) STARCH. (B) ADD WATER TO A MIXTURE OF *KUDZU* STARCH, TSX, AND WHITE SUGAR. (C) HEAT THE MIXTURE OVER MEDIUM HEAT WHILE STIRRING. AFTER BOILING, LOWER THE HEAT. (D) CONTINUE TO HEAT THE MIXTURE AND STIR UNTIL IT BECOMES TRANSPARENT. (E) COOL THE *KUDZU-MOCHI* UNTIL IT SETS. (F) MIX BROWN SUGAR AND WATER IN A PAN AND HEAT UNTIL THICKENED. (G) CUT THE *KUDZU-MOCHI* INTO BITE-SIZED PIECES, POUR THE BROWN SUGAR SYRUP OVER THEM, AND SPRINKLE WITH SOYBEAN FLOUR.

3. Pour the *kudzu-mochi* (2) into a moistened rectangular mold, flatten the surface, and cool it down to <20°C in the refrigerator or on an ice-water bath until it sets (Figure 19.3E).
4. Put brown sugar and 50 mL water in a pan and heat for 5 minutes until it thickens (100°C); cool in the refrigerator or on an ice-water bath (Figure 19.3F).
5. Cut the *kudzu-mochi* (3) into bite-sized pieces, place them on a dish, pour the brown sugar syrup (4) over them, and sprinkle with soybean flour (Figure 19.3G).

pH = 7.04

Preparation hints:

19.6.3.1 Stir well before heating so that lumps do not form.

19.6.3.2 Transparent jelly will form on the bottom and sides of the pan; stir and knead well and vigorously because gelatinization occurs suddenly.

This recipe is popular on hot, summer days in Japan. However, today, *kudzu-mochi* is rarely eaten in Japan because arrowroot starch is expensive; sweet potato starch is used as a less expensive substitute and the jelly is called *warabi-mochi*. TSX is used as a gelling agent and an inhibitor of starch gelatinization in this recipe.

19.6.4 Sweet Red Bean Soup (*Shiruko*) (Figure 19.4)

(Serves 5)

1.5 g (½ tsp) TSX powder

2.0 g (1 tsp) κ-carrageenan powder

350 mL (1⅖ cups) water

200 g red bean paste

120 g (a little under 1 cup) white sugar

Salt

Rice cake (optional)

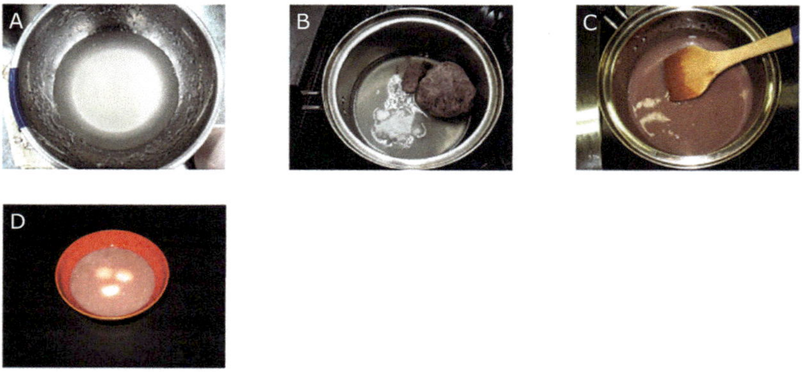

FIGURE 19.4 SWEET RED BEAN SOUP (*SHIRUKO*). (A) MIX WATER, TSX, AND CARRAGEENAN. (B) ADD RED BEAN PASTE TO THE TSX SOLUTION. (C) ADD WHITE SUGAR AND SALT AND HEAT. (D) POUR THE SWEET RED BEAN SOUP INTO A BOWL AND ADD RICE CAKE (OPTIONAL).

1. Put water in a bowl and stir in TSX and carrageenan with a hand mixer (Figure 19.4A).
2. Pour the TSX solution (1) into a pan and add red bean paste (Figure 19.4B), white sugar, and a pinch of salt (Hint 19.6.4.1) and heat (Figure 19.4C).
3. Pour the sweet red bean soup (2) in a soup bowl and add rice cake (optional) (Figure 19.4D).

pH = 6.65

Preparation hints:

19.6.4.1 Adding a small quantity of salt to sweet food increases its sweetness. In Japan, *shiruko* is served with vegetables or seaweed brined in salt.

This recipe is very popular in Japan, and is good for cold, winter days. TSX and carrageenan are added as suspension-stability agents. Figure 19.5 shows *shiruko* with (left) and without (right) TSX and carrageenan, 10 minutes after it was poured into glasses.

FIGURE 19.5 *SHIRUKO* WITH (LEFT) AND WITHOUT (RIGHT) TSX AND CARRAGEENAN, 10 MINUTES AFTER IT IS POURED INTO GLASSES.

19.7 TIPS FOR THE AMATEUR COOK AND PROFESSIONAL CHEF

- The TSX solution is resistant to a wide range of pH values, heating, freeze-thaw cycles, and salts.
- Dissolving TSX in water is very important to obtain the effect of increased viscosity or gelling. To dissolve TSX, a large volume of water is required, and the mixture should be agitated at high speed.
- Adding 40% sugar to jelly with TSX results in an elastic jelly that does not exhibit syneresis.
- The melting temperature of jelly with TSX is 20°C–30°C.
- Adding polyphenols from green tea, for example, to jelly made with TSX, results in a stronger and more brittle gel.

REFERENCES AND FURTHER READING

Hayashi, T. 1989. Xyloglucans in the primary cell wall. *Annu. Rev. Plant Physiol. Plant Mol. Biol.* 40:139–68.

de Lima, D. U., and M. S. Buckeridge. 2001. Interaction between cellulose and storage xyloglucans: the influence of the degree of galactosylation. *Carbohydr. Polym.* 46:157–63.

Maeda, T., Yamashita, H., and N. Morita. 2007. Application of xyloglucan to improve the gluten membrane on breadmaking. *Carbohydr. Polym.* 68:658–64.

Nishinari, K., Yamatoya, K., and M. Shirakawa. 2000. Xyloglucan. In *Handbook of hydrocolloids*, eds. G. O. Phillips and P. A. Williams, 247–67. Cambridge: Woodhead Publishing Limited.

Nussinovitch, A. 1997. *Hydrocolloid applications. Gum technology in the food and other industries*. London: Blackie Academic & Professional.

Nussinovitch, A. 2003. *Water soluble polymer applications in foods*. Oxford: Blackwell Publishing.

Nussinovitch, A. 2010. *Plant gum exudates of the world sources, distribution, properties and applications*. Boca Raton, FL: CRC Press, Taylor & Francis Group.

Nussinovitch, A., and M. Hirashima. 2014. *Cooking innovations, using hydrocolloids for thickening, gelling and emulsification*. Boca Raton, FL: CRC Press, Taylor & Francis Group.

Shirakawa, M., Yamatoya, K., and K. Nishinari. 1998. Tailoring of xyloglucan properties using an enzyme. *Food Hydrocolloids* 12:25–8.

Simi, C. K., and T. E. Abraham. 2010. Transparent xyloglucan–chitosan complex hydrogels for different applications. *Food Hydrocolloids* 24:72–80.

Takeda, T., Furuta, Y., Awano, T., Mizuno, K., Mitsuishi, Y., and T. Hayashi. 2002. Suppression and acceleration of cell elongation by integration of xyloglucans in pea stem segments. *Proc. Natl. Acad. Sci. USA* 99:9055–60.

Temsiripong, T., Pongsawatmanit, R., Ikeda, S., and K. Nishinari. 2005. Influence of xyloglucan on gelatinization and retrogradation of tapioca starch. *Food Hydrocolloids* 19:1054–63.

Yamanaka, S., Yuguchi, Y., Urakawa, H., and K. Kajiwara. 1999. Gelation of enzymatically degraded xyloglucan extracted form tamarind seed. *Sen'i Gakkaishi* 55: 528–32.

Yamanaka, S., Yuguchi, Y., Urakawa, H., Kajiwara, K., Shirakawa, M., and K. Yamatoya. 2000. Gelation of tamarind seed polysaccharide xyloglucan in the presence of ethanol. *Food Hydrocolloids* 14:125–8.

York, W. S., van Halbeek, H., Darvill, A. G., and P. Albersheim. 1990. Structure analysis of xyloglucan oligosaccharides by 1H-NMR spectroscopy and fast-atom-bombardment mass spectroscopy. *Carbohydr. Res.* 200:9–31.

Yoshimura, M., Takaya, T., and K. Nishinari. 1999. Effects of xyloglucan on the gelatinization and retrogradation of corn starch as studied by rheology and differential scanning calorimetry. *Food Hydrocolloids* 13:101–11.

Yuguchi, Y. 2002. The various types of gelation mechanism of xyloglucan. *Cellulose Comm.* 9:76–80.

Yuguchi, Y., Mimura, M., Urakawa, H., Kajiwara, K., Shirakawa, M., Yamatoya, K., and S. Kitamura. 1996. Crosslinking structure formation of some polysaccharides in aqueoussolution. *Proceedings of the international workshop on green polymers*, Bandung-Bogor, Indonesia, 306–29.

Yuguchi, Y., Urakawa, H., Kajiwara, K., Shirakawa, M., and K. Yamatoya. 2000. Gelation of xyloglucan polysaccharide extracted from tamarind seed. *Proceedings of the second international workshop on green polymers*, Bandung-Bogor, Indonesia, 253–63.

CHAPTER 20

Future Ideas for Hydrocolloid Processing and Cooking

20.1 INTRODUCTION

The global hydrocolloid market reached a value of $6.6 billion in 2015 [equivalent to 1.66 million metric tons (MT)]. Growing at an estimated 5-year (2015–2020) compound annual growth rate (CAGR) of 4.4%, the market should reach 2.1 million MT by 2020, with a value of almost $8.2 billion. The introduction of novel hydrocolloids in food processing constitutes a great business incentive to gain the obligatory legislative approval. Furthermore, some specific hydrocolloids have a long history of use in food applications in the Far East, and will, undoubtedly, be introduced to the West in the near future as food additives. The quest for synergistic mixtures and our understanding of such interactions will help in developing innovative processing procedures and exclusive cooking innovations.

Another aspect that will progressively become more significant is the nutritional value of these hydrocolloids, such as the ability to decrease blood cholesterol levels and prebiotic effects. These are just a few benefits of hydrocolloids that will undoubtedly contribute to their increased future consumption in food and cooking. In brief, a clean label (vegetable origin), health benefits (fat reduction), and cost reduction are some of the incentives that have led to global efforts in hydrocolloid innovations. It is anticipated that in the coming years these factors will become increasingly important in hydrocolloid innovations.

20.2 FUTURE TRENDS FOR SEVERAL GELLING AGENTS AND VISCOSITY FORMERS

Arabinoxylan (Chapter 2) is a major fiber. The water-extractable constituent of arabinoxylan displays high viscosity when it is dispersed in water. As a food component, arabinoxylan can affect water-holding ability, dough rheology, and starch retrogradation.

Addition of 0.25%–1.0% arabinoxylan to wheat flour increases the quality of Chinese noodles. The water absorption capacity of the noodles increases and the cooking losses are reduced. The rate of amylopectin crystallization in starch gels containing arabinoxylans is either increased or decreased, depending on the final water content of the starch. Substitution of 50% durum wheat semolina with pearling byproducts results in a darker pasta with good cooking features, related to stickiness, firmness, and cooking losses. The inclusion of arabinoxylans gives these products a high dietary content, making them a healthy alternative for health-conscious consumers. Arabinoxylans have significant effects on cereal processes, such as milling, brewing, and bread-making. They can also be used as film-forming agents, cryostabilizers, and surface-active agents in numerous food products.

Cereal β-glucans (Chapter 4) are hydrocolloid polysaccharides found in cereals, and predominantly in oats and barley at high concentrations. As new technologies get developed for the manufacture of these ingredients, their inclusion in food products is expected to increase in the future, particularly in those food products where fiber is required. Beyond their function of thermal gelation, the group of cellulose derivatives covered by methylcellulose/hydroxypropyl methylcellulose (MC/HPMC) also confers a comprehensive variety of textural properties to food items. Cellulose derivatives will play a vital part in the development of no-fat and low-fat substitutes for traditional salad dressings. These derivatives can be classified as novel products, which require water control in multiphase systems. These derivatives also form novel freeze-thaw-stable products, where they will enable texture retention during the prolonged shelf life. The future uses of microcrystalline cellulose (MCC) are also related to the development, advancement, and applications in the health sector. Colloidal MCC-based lipid compositions might be useful in convenience cooking sauces, spreads, and other confections. A MCC-based bulking agent can also be used for calorie reduction in future cooking and food products. The novel bacterial cellulose (Chapter 3), distinguished for its unique fiber dimensions and large surface area, can be employed to form a thermally stable colloidal network with unsurpassed suspension properties. Addition of 10% bacterial cellulose to Chinese-style meatballs results in a product with acceptable textural and sensory qualities. Inclusion of a different form of colloidal MCC in an ice cream mix has reportedly produced a light and lower-fat ice cream that has the sensory perception of the full-fat recipes. Other new uses for MCC include their addition to fillings and sugary confections in low-moisture systems to retain the texture and sensory properties. The gum formed by the combination of MCC and *konjac-mannan* can be useful as a unique stabilizer.

Chitosan (Chapter 5) is a natural, biodegradable, and biocompatible polysaccharide derived from the deacetylation of chitin (a polysaccharide found in the exoskeleton of crustaceans and insects). It is the second most abundant natural polysaccharide found on earth (cellulose being the first). Chitosan has been widely investigated as a natural biomaterial for many biomedical applications due to its biocompatibility, biodegradability, antimicrobial properties, and functionality. Chitosan promises good performance in the area of dietary food. In yet another application, chitosan and kieselsol are often sold as a set, in sealed liquid envelopes, for wine fining, with chitosan being fairly gentle on the character of the finished wine. It is expected that further exploration of the chemistry of chitosan-based gels will lead to the formation of similar gels that will help in expanding the current range of applications in the near future.

Xanthan gum has become one of the most cost-competitive multipurpose hydrocolloids. It has a somewhat multifaceted molecular structure, but natural deviations from the ideal structure can be found as well. An understanding and handling of these difference can lead to new and advanced functionality in the food industry and in the kitchen. Theoretically, additional microbial polysaccharides, i.e., pullulan (Chapter 16), scleroglucan, elsinan (Chapter 15), levan (Chapter 12), and dextran (Chapter 6), have potentially significant uses in the food industry and in cooking, but not all of them have clear global approval as food additives.

20.3 FUTURE TRENDS IN PROTEIN HYDROCOLLOIDS

A large number of milk proteins (Chapter 14) still continue to be recovered from milk. These can be utilized in both the industry and the kitchen. Furthermore, tailoring milk-protein products through physical and enzymatic modifications can lead to their usage in the novel field of milk-protein constituents, for explicit food applications. Only a small number of active peptides derived from milk proteins are commercially obtainable. When these peptides interact with different food ingredients, they influence the biological activity of the food, and this phenomenon can be tapped in novel processing techniques and possibly new nutritional and nutraceutical applications. Vegetable protein isolates (Chapter 18) are important as functional ingredients in various food formulations. A better understanding of their functional properties can contribute to their inclusion in foams, emulsions, gels, and doughs for texturization and enhancement of fat- and water-holding capacities.

20.4 UNIQUE NUTRITIONAL AND FUTURE HEALTH CLAIMS OF HYDROCOLLOIDS

20.4.1 Dietary Fibers

Dietary fiber offers a range of health benefits. The most significant among such benefits is a possible decrease in the risk of chronic diseases, such as cardiovascular disease, diabetes, diverticulitis, and obesity. Elevating the quantity and varying the structure of arabinoxylans (Chapter 2) in the endosperm can decrease the glycemic impact and boost the prebiotics of rice and other cereals. Remodeling the fine structure of arabinoxylan by the use of arabinoxyl anarabinofuranohydrolase (AXAH), which is involved in the elimination of arabinosyl residues, is a feasible approach. It has been suggested that a high degree of arabinosyl substitution in the xylan backbone increases its solubility and, consequently, its nutritional value as a soluble dietary fiber.

Bacterial cellulose (Chapter 3) is a dietary fiber classified as generally recognized as safe (GRAS). It has some advantages as compared to other dietary fibers. It is a very pure form of cellulose and, in contrast to cellulose obtained from plant sources, it does not require harsh chemical treatments for separation and purification. In addition, bacterial cellulose cultured in fruit syrup can take on both the flavor and the pigments of the syrup. Bacterial cellulose can produce a variety of textures, shapes, and products and, consequently, it may have diverse applications in food and cooking. Furthermore, the fibers of bacterial cellulose have a fine 3D-network structure; their size, in the order of nanometers, permits their use in novel industrial procedures for manufacturing and processing food.

20.4.2 Health Claims and Related Issues

The inherent mixed-linkage β-glucans derived from cereals (Chapter 4) are classified as soluble dietary fiber, and they have rheological properties that are comparable to those of other randomly coiled polysaccharides and guar gum. The ability of barley and oat food items to weaken the insulinemic response and postprandial glycemia is related to the presence of β-1,3 or β-1,4 linkages and to viscosity. However, the role of β-glucan viscosity in lowering the serum cholesterol levels has not yet been directly validated. Furthermore, not all studies report a statistically significant lowering. On the other hand, the extensive range of reported effectiveness may be explained, in part, by the properties of the β-glucans that are used in the diet, in conjunction with its dose.

Dextran (Chapter 6) is a polymer of dextrose. It is synthesized from sucrose by *Leuconostoc mesenteroides*. A number of microorganisms produce dextrans, which vary in their molecular weights and have structures with varying degrees of branching. Dextran produces low-viscosity solutions, which differentiates it from other high-molecular-weight polysaccharides. It is a neutral polymer. Dextrans have an inhibitory effect on thrombocyte aggregation and coagulation factors and they are used as volume expanders in pregnancy during emergency situations. Iron dextran is a form of the mineral iron, which is essential for organisms, particularly for the transport of oxygen in the blood. Iron dextran is utilized to treat iron deficiencies and iron deficiency anemia (low red blood cells). Nevertheless, possible side effects of iron dextran include allergic reactions that cause hives, breathing problems, and swelling of face, lips, tongue, or throat. Iron dextran can also cause severe and, sometimes fatal, allergic reactions or critically low blood pressure.

Gum ghatti (Chapter 7) is an exudate secreted from the wounds in the bark of *Anogeissus latifolia,* a large tree located in the forests of Sri Lanka and India. These exudates have been comprehensively used not only as a medicinal additive, such as a coating agent for tablets, but also as an emulsifier and stabilizer in beverages and food products; they also have additional industrial applications, such as in paints and ink. Gum ghatti was originally used as an alternative to gum arabic owing to their comparable properties. Gum ghatti has been utilized as a chemical addition to food products owing to its outstanding emulsification properties. Another exudate, gum karaya (Chapter 8), is derived from *Sterculiaurens* Roxb. and other *Sterculia* spp. (fam. Sterculiaceae). It is a complex, partially acetylated, polysaccharide of very high molecular weight. Gum karaya (Chapter 8) is used as a thickener in cosmetics, denture adhesives, and various medications. In food and beverages, it is regarded as a binder. Gum karaya is used as a bulk-forming laxative to relieve constipation that occurs as a result of swelling in the intestine and it also stimulates the digestive tract. Gum tragacanth (Chapter 9) is another natural exudate obtained from the dried sap of several species of legumes of the genus *Astragalus,* found in the Middle East. Gum tragacanth can be regarded as a natural form of soluble dietary fiber, which has a long history of safe use in food and pharmaceutical formulations. Gum tragacanth can lead to the progress of innovative functional food with health claims. Furthermore, it can serve as an additional foundation for the manufacture of prebiotics or prebiotic dietary fibers, like arabinogalactan, xylogalacturonan, and fuco-xylogalacturonan polymers, because of its exceptional sugar composition and structure.

Inulin (Chapter 10) is a common term that includes all linear fructans with β-(2,1) fructosyl-fructose glycosidic bonds. Inulin can be found in various types of vegetation, and it can also be produced by microorganisms. It is found in banana, barley, chicory, onion, and wheat, among many other sources. Inulin and fructo-oligosaccharides (FOS) are associated products that are based on fructose. These dietary fibers ferment rapidly upon entering the colon. Used together, inulin and FOS have been shown to encourage bifidobacteria and lactobacillus, thereby satisfying the definition of a prebiotic. Stimulation of bifidobacteria was demonstrated when 5–15 g of inulin and FOS was consumed per day. Inulin and FOS can be used for fiber fortification in beverages, for example, carbonated and non-carbonated drinks, dairy and dairy-replacement beverages, dry mixtures, near-water beverages, juices, energy and sport drinks, coffee, tea, and water. In brief, inulin, individually or in combination with FOS, can contribute to the production of low-calorie foodstuffs, a health claim that is particularly relevant in view of the obesity pandemic.

Levans (Chapter 12) are natural polymers of fructose and are found in numerous microbes and plants. Comparable to dextrans, they are a byproduct of sugared juice production. On the other hand, levans, which can be simply manufactured from sucrose, have possible industrial applications as viscosity-formers and encapsulating agents. They can also deliver supplementary, desired products from sugar cane juice. The hypocholesterolemic effect of levans may result from the inhibition of intestinal sterol absorption, rather than from the action of their fermentation products. A levan-type exopolysaccharide (EPS) derived from *Paenibacillus polymyxa* was effectively acetylated, phosphorylated, and benzylated, to give acetylated levan, phosphorylated levan, and benzylated levan, respectively. The in vitro antioxidant and antitumor activities of the natural polysaccharide and its derivatives were defined. In comparison to the natural polysaccharide, EPS-1, the acetylated levan, benzylated levan, and phosphorylated levan all show better reducing power and better scavenging activity on hydroxyl radicals than on superoxide radicals. Furthermore, acetylated levan, phosphorylated levan, and benzylated levan display higher antiproliferative activity against human gastric cancer cells in vitro. The enhanced activities of these derivatives are probably due to the introduction of acetyl, benzyl, or phosphoryl groups into the EPS-1 molecule, which enhance the electron-donating capability and the affinity for receptors on immune cells. As a consequence, the derivatives may serve as promising antioxidant and antitumor agents.

Mesquite, or *Prosopis*, gum (Chapter 13) has been extensively used by the Indian cultures of central northwestern Mexico and the southwestern United States, since pre-Columbian times. It is mostly used as a candy, as a component in animal and human foodstuffs, as a therapeutic assistance for sore eyes, uncomfortable throat, stomach pain, and diarrhea, for preventing infections, and for the treatment of open wounds. Mesquite gum has been comprehensively used in minor processing industries, mostly dealing with food and confections. It has healing qualities because of its natural mucilage content. It is believed to soothe stomach/intestinal pain and other abdominal problems, and it also serves as an eye soother and to ease respiratory problems, but further studies must be performed to verify these claims.

Pullulan (Chapter 16) is a neutral polymer composed of α-1,6-linked maltotriose residues, which in turn are comprised of three glucose molecules linked to each other by an α-1,4 glycosidic bond. Pullulan is a film former and binder that is commonly used in the manufacture of breath strips. It is also used as a component in beauty products and cosmetics due to its water solubility and adhesive properties. Its preliminary use in dietary supplements and health food applications will be expanded in the near future to include pharmaceutical applications. The pullulan is recognized as a dietary fiber as well as an immune stimulator. It also has special effects on health because of its function as a dietary fiber and its immunostimulatory activity.

Commercial soluble soybean polysaccharide (Chapter 17) is extracted from fibrous bean curd residue, which is the byproduct of *tofu*, soymilk, and soybean protein isolate. Soy fiber has been reported to decrease blood cholesterol, improve laxation, and decrease the risk of diabetes. As dietary fiber consumption is lower than recommended for many people, strategies are needed to effectively fortify nutrition with fiber. It is possible to manufacture dairy products fortified with soluble soybean polysaccharide (beverages, puddings, and low-fat ice cream), that will satisfy customers with a textural range comparable to that of the existing products.

Vegetable protein isolates (Chapter 18) might serve as functional ingredients in food formulations. Their main sources are legumes, cereals, oilseeds, roots, and green leaves, though there is a distinct predominance in soybean, pea, and wheat. More than a few approaches have been established to increase the nutritional importance of vegetable proteins. Most frequently used are seed dehulling, irradiation, protein fractionation, soaking, and thermal treatments. Additional methods are based on breeding, germination, hydrolysis, and fermentation.

20.4.3 Alternatives to Hydrocolloids

The supply of gum arabic, which is produced by acacia trees in the gum belt of Africa, is unpredictable due to political and climatic issues in the main producing countries, such as Sudan and Nigeria. This has led to spikes in the price of this ingredient. Nevertheless, gum arabic has been recognized over the last few years as a stable and well-tolerated soluble fiber, with proven prebiotic and hypoglycemic effects. Maltodextrin and whey protein isolates (Chapter 14) can produce complexes with exceptional emulsification properties, and they can offer users an alternative to gum arabic, particularly for applications in soft drinks. These systems can be effectively diluted with carbonated sugar syrup to generate stable, dilute, and colored emulsions. Another gum arabic substitute might be an enzyme-modified soluble soybean polysaccharide (Chapter 17), which can be used in flavor emulsions. Some innovative hydrocolloid developments include hydrocolloid replacements, for example, gelatin replacement by MC or HPMC. On one hand, gelatin is a multifunctional ingredient that is compatible with most materials. It may therefore play a role in the development of novel sugar confectionery products. On the other hand, substitution of gelatin with some other hydrocolloid might be desired to get a kosher/halal/vegetarian label. Collagen hydrolysate has been mentioned, especially for chocolate applications, as it can replace some of the cocoa butter, with minimum impact on chocolate viscosity, moisture, taste and, sensory evaluation.

20.5 CONCLUSION

Hydrocolloids offer dozens, if not hundreds, of modifications and alternatives. The large number of available hydrocolloids that can formulate consumable items creates a challenge for food formulators. Potential developments will certainly concentrate on the diet, physical conditions, and wellness of the consumers.

References and Further Reading

Dickinson, E. 1991. *Food polymers, colloids, gels and colloids.* Cambridge: Royal Society of Chemistry.

Dickinson, E., and T. van Valiet. 2003. *Food colloids: biopolymers and materials.* Cambridge: Royal Society of Chemistry.

Fan, L., Ma, S., Wang, X., and X. Zheng. 2016. Improvement of Chinese noodle quality by supplementation with arabinoxylans from wheat bran. *Int. J. Food Sci. Technol.* 51:602–8.

Harris, P. 1990. *Food gels.* London and New York: Elsevier Science Publishers Ltd.

Imeson, A. 1992. *Thickeners and gelling agents for food.* London: Blackie Academic and Professional.

Imeson, A. 2010. *Food stabilizers, thickeners and gelling agents.* Oxford: Wiley-Blackwell, John Wiley & Sons Ltd. publication.

Nussinovitch, A. 1997. *Hydrocolloid applications. Gum technology in the food and other industries.* London: Blackie Academic and Professional.

Nussinovitch, A. 2003. *Water-soluble polymer applications in foods.* Oxford: Blackie Academic and Professional.

Nussinovitch, A. 2010. *Polymer macro- and micro-gel beads*. New York: Springer.
Nussinovitch, A. 2010. *Plant gum exudates of the world: sources, distribution, properties and applications*. Boca Raton: CRC Press, Taylor and Francis Group.
Nussinovitch, A., and M. Hirashima. 2014. *Cooking innovations, using hydrocolloids for thickening, gelling and emulsification*. Boca Raton, FL: CRC Press, Taylor & Francis Group.
Phillips, G. O., and P. A. Williams. 2009. *Handbook of hydrocolloids*. 2ndedn. Oxford: Woodhead Publishing Limited.
Shi, Z., Zhang, Y., Phillips, G. O., and G. Yang. 2014. Utilization of bacterial cellulose in food. *Food Hydrocolloids* 35:539–54.
Wood, P. J. 2007. Cereal β-glucans in diet and health. *J. Cereal Sci.* 46:230–8.
http://www.foodnavigator.com/Science-Nutrition/Enzyme-modified-soybean-sugar-another-gum-arabic-replacer.
http://www.foodingredientsfirst.com/news/Hydrocolloid-Innovation-Shortlist-Revealed.htm.
http://www.marketwired.com/press-release/diverse-industries-thickening-global-hydrocolloids-market-reports-bcc-research-2100914.htm.
https://www.sciencedirect.com/topics/agricultural-and-biological-sciences/arabinoxylan.
http://womenatbayview.com/health-library/hw-view.php?DOCHWID=d03767a1.
http://www.teknoscienze.com/agro/pdf/GAVLIGHI_AF2_2013.pdf.
https://www.henriettes-herb.com/archives/best/2000/mesquite.html.
https://www.nature.com/articles/s41598-017-03053–9.

GLOSSARY

Baking powder: A dry chemical leavening agent. A mixture of a carbonate or sodium bicarbonate and a weak acid. It is used in baking for increasing the volume and lightening the texture of baked goods.

Baking soda: Sodium bicarbonate used in baking as a leavening agent. It has a slightly salty, alkaline taste resembling that of washing soda.

Bamboo shoot: Bamboo sprout or young shoot. Eaten as a vegetable in the spring, in East Asia, Southeast Asia, and India.

Bittern: A coagulant for *tofu*. An aqueous solution containing ~0.4% magnesium chloride.

Blender: Electrically-powered standing mixer. Used to mix, purée, emulsify, etc.

Bread machine: A home appliance for making bread, invented in Japan.

Buttermilk: Liquid left over from churning butter.

Cane syrup: A thick, aromatic syrup made from sugarcane juice, containing refined sugar.

Celery seed: Used as a spice or herb. Also used as a medicinal treatment for pain relief.

Citric acid: An organic acid found mainly in citrus fruit, and various other kinds of fruit and vegetables.

Clove: The flower bud of the *Myrtaceae* tree. Used in the dried form, as a flavoring spice.

Cornmeal: Coarse flour from ground dried corn.

Cumin seed: Used in the dried form, as a flavoring spice.

Cup: In this book, 1 cup = 250 mL. The volume of a US cup is 240 mL, so if you are using a US cup you need to adjust the volume a little bit. Similarly, a Japanese cup = 200 mL, so 1 cup in this book is equivalent to 5/4 Japanese cups.

Dried bonito: A fermented fish product called *katsuobushi* in Japan. An extremely hard food that is shaved into flakes for use. It has an *umami* taste and is used in Japanese broth.

Durum semolina flour: Also called pasta wheat flour or macaroni wheat flour, ground from hard wheat.

Glacial acetic acid: An organic acid found mainly in vinegar. Vinegar contains about 5% acetic acid.

Glucose: A type of sugar and one of the monosaccharides, termed as blood sugar. It has 60%–80% of the sweetness of sucrose. It is the only energy source used by our brains.

Glutinous: A grain type. Its starch consists of 100% amylopectin. Rice cakes are made from glutinous rice.

Guar gum: A polysaccharide extracted from leguminous (guar) plants. It is one of the galactomannans, and is used as a food additive, most widely in milk products.

Hand mixer: Electrically-powered hand-held mixer. Usually has one or two beaters and is used to mix, stir, beat, whip, etc.

Ice cream maker: Machine to make small quantities of ice cream at home. Also called an ice cream machine or an ice cream freezer.

Japanese broth: Kelp (*Laminaria* spp.) and dried bonito are usually used for making a Japanese broth called *dashi*. The recipe follows:

Ingredients: 10 g kelp, 10 g dried bonito flakes, 1000 mL water.

1. Soak kelp in water in a pan for 30 minutes, and then heat to boiling.
2. Remove kelp from the pan and add dried bonito; boil for 1–2 minutes and remove from heat.
3. Strain the broth.

Japonica rice: Short-grained subspecies of rice. Comes in non-glutinous and glutinous forms.

***Kimchi*:** Korean fermented pickled food. Chinese cabbage, Japanese white radish, and cucumber, among others, can be used.

Kirshwasser: Kirsh or cherry brandy. Also used as a flavoring agent for sweets.

***Konjac*:** A Japanese gelling foodstuff made from *konjac* flour.

Laver: A kind of seaweed. As a food, it is prepared by spreading seaweed paste and drying. It is widely eaten in East Asia.

Lecithin: A phospholipid used as an emulsifier. Egg yolk and soybeans contain large amounts of lecithin.

Lotus root: Root of the lotus plant, used as a vegetable in Asian cuisines.

Maple syrup: Syrup from the xylem sap of the sugar maple.

***Mirin*:** Sweet alcohol made from rice in Japan. Contains about 14% alcohol.

Monosodium glutamate: An *umami* seasoning. Widely used in East Asian and Southeast Asian cuisines.

Mother liquor: Solution left over, after crystallization in sugar refining.

Non-glutinous: A grain type. Its starch consists of about 20% amylose and 80% amylopectin. Non-glutinous rice is eaten daily as steamed rice.

***Okara*:** Soybean bran or pulp (also known as *tofu* lees). A byproduct of the tofu-making process. It is not tasty, but contains a great deal of fiber.

***Piment d'Espelette*:** Sweet pepper cultivated in Espelette, France. Used in traditional Basque dishes.

***Ragi tempeh*:** A type of bacteria used to ferment *tempeh*.

Red bean: *Azuki* bean used in China, Japan, and Korea.

Rice cooker: An automatic machine for steaming rice. Both electric and gas models are available, but the electric one is more widely used in Japan.

***Sake*:** Rice wine containing about 15% alcohol.

***Shiitake* mushroom:** A representative of the Japanese mushrooms. It has a distinct *umami* flavor.

Shortening: A type of fat which was developed as a substitute for lard.

Small dried fish: Dried sardines or dried anchovies. Used to make Japanese and Korean soup stocks. The recipe follows:
Ingredients: 30 g small dried fish, 1000 mL water.

1. Remove the head and insides of the dried fish and soak in a pan for at least 30 minutes; heat to boiling, reduce the heat to low, and continue to simmer for a few minutes.
2. Strain the broth.

Sour cream: A product in daily use, consisting of fermented heavy cream.

Soy sauce: A basic seasoning in Japanese dishes made from soybeans, wheat, and salt. Heavy soy sauce, which is termed "soy sauce" in this book, is the most widely used variant, although light soy sauce is also used in cooking.

Tablespoon: 1 tablespoon = 15 mL. The abbreviation for tablespoon is "tbsp."

Teaspoon: 1 teaspoon = 5 mL. The abbreviation for teaspoon is "tsp."

Tomato purée: A type of processed tomato food. Usually consists of only tomato.

Trisodium citrate: Known as sour salt. Used as a flavoring agent and preservative.

Whisk: Hand-operated mixer. Used to mix, stir, beat, whip, etc.

Xanthan gum: A polysaccharide manufactured by biotechnological processes. A food additive that is widely used as a thickening agent and stabilizer.

INDEX

H

L

M

R

S

T

U

V

Y

Z